AutoCAD® R14 Fundamentals

KNOWLEDGEWORKS™

Autodesk.
Press

I(T)P® An International Thomson Publishing Company

Albany • Bonn • Boston • Cincinnati • Detroit • London • Madrid
Melbourne • Mexico City • New York • Pacific Grove • Paris • San Francisco
Singapore • Tokyo • Toronto • Washington

NOTICE TO THE READER

Trademarks

AutoCAD®, Autodesk, and the Autodesk logo are registered trademarks of Autodesk, Inc. Microsoft, Windows, and the Windows logo are registered trademarks of Microsoft Corporation. KnowledgeWorks is a trademark of HTR, Inc. All other product names are acknowledged as trademarks of their respective owners.

COPYRIGHT © 1998
By HTR, Inc.

For more information, contact:

Autodesk Press
3 Columbia Circle, Box 15-015
Albany, New York USA 12212-5015

International Thomson Editores
Campos Eliseos 385, Piso 7
Colonia Polanco
11560 Mexico D. F. Mexico

International Thomson Publishing Europe
Berkshire House 168-173
High Holborn
London, WC1V 7AA
United Kingdom

International Thomson Publishing GmbH
Konigswinterer Strasse 418
53227 Bonn Germany

Thomas Nelson Australia
102 Dodds Street
South Melbourne, Victoria 3205
Australia

International Thomson Publishing France
Tour Maine-Montparnasse
33, Avenue du Maine
75755 Paris Cedex 15, France

Nelson Canada
1120 Birchmont Road
Scarborough, Ontario
Canada, M1K 5G4

International Thomson Publishing--Japan
Hirakawacho Kyowa Building, 3F
2-2-1 Hirakawa-cho Chiyoda-ku
Tokyo 102 Japan

International Thomson Publishing Southern Africa
Building 18, Constantia Park
240 Old Pretoria Road
P.O. Box 2459
Halfway House, 1685 South Africa

International Thomson Publishing Asia
221 Henderson Road
#05-10 Henderson Building
Singapore 0315

3 4 5 6 7 8 9 10 XXX 01 00 99

ISBN 0-7668-0219-1

PRINTED IN CANADA

Quick Table of Contents

Table of Contents

Chapter 1: Introduction to AutoCAD R14

Chapter 24: Using Utility Commands

Introduction

Welcome to AutoCAD R14 Fundamentals

This book was developed to provide training on the basic concepts of Autodesk's AutoCAD Release 14 software. The information and exercises presented here are intended to provide AutoCAD users with both an *understanding* of how to use these basic features and with practical *hands-on exercises*.

Like earlier versions of AutoCAD, Release 14 provides new features, functionality, and improvements that enable you to work *faster, more easily, and with higher quality results*. This book provides an excellent overview and many practical exercises on the basic concepts and features found in Release 14. You will find some topics more appropriate to your immediate needs than others, and perhaps some that you will not use at all. Certain topics require networked labs or additional hardware to which you may not have access.

Objectives

Upon completion of this book, you will be able to:

▶ Start AutoCAD; navigate the User Interface, access menus, dialog boxes, toolbars, and on-line help; and set parameters for a drawing session

▶ Create new drawing files, open existing drawings, save changes to existing drawings, and exit AutoCAD.

▶ Use a template to create new drawing files that contains standard settings, customize AutoCAD drawing settings, calculate the size of the drawing area, and use drawing aids.

▶ Use layers to control the color, linetype and visibility of objects in drawings.

▶ Use display commands, and control the view of objects displayed in the drawing window.

▶ Use AutoCAD Draw commands to create objects that can be edited or modified.

▶ Use Object Snaps to reference specific locations on selected objects, and ensure quality and accuracy in the drawing process.

▶ Use AutoCAD Modify commands to simplify the drawing process, increase drawing productivity, logically structure and avoid repetitive drawing.

◗ Use Inquiry commands to provide database information for selected objects, display the status of a drawing, and document the time that spent working on the drawing.

◗ Use a of number Annotation techniques to present text, dimensions, notes, and titles to drawings.

◗ Place dimensions in drawings that conform to local and national standards.

◗ Use blocks to combine, organize and manipulate several objects as a single object, and use them to reduce drawing file size.

◗ Use the PURGE and RENAME utility commands to reduce file sizes, and rename objects.

◗ Use the Print/Plot Configuration dialog box to improve the appearance and efficiency of final plotted output.

Who Should Read This Book

This course is designed for the new or novice AutoCAD user as well as DOS-based AutoCAD user. It is assumed that the reader will possess the following skills and abilities:

◗ General knowledge of the use of a personal computer

◗ General knowledge of design and drafting concepts is helpful, but not necessary

Style Conventions Used in this Book

The following table describes the style conventions used throughout this book:

Style Convention	Type of Text and Example
Bold	Keyboard-entered text: **stretch** Coordinate points: **3,2** Names of axes: **X,Y,Z axis**
Italics	Instructions following the prompt: *line from prompt* Names of books used as references: *Microsoft Word for Windows 7.0* For emphasizing a word or phrase: This version enables you to solve problems *faster*. Path Names: *myfile/support/samples* File names, stand-alone file extensions, directory names, drawing names: *template.dot* Items in a bulleted list followed by descriptions: *Item* - Description of the item
SMALL CAPITALS	Product commands: RTZOOM, BASE Environment variables: ACADMAXMEM System variables: VISRETAIN
ALL CAPITALS	Names of keys on the keyboard: ENTER, SHIFT, CTRL
Title Style Capitalization	Dialog box names: Boundary Hatch dialog box Areas: The Description area Options: Move option Menu names: File menu Toolbar names: Draw toolbar Titles of chapters, parts, books: *AutoCAD Release 13 User's Guide*
all lowercase	File names: *template.dot* Stand-alone file extensions: *.dwg, .dot* Path names: *myfile/support/samples* Object names: line, circle, ellipse Drawing names: *drawing.dwg*
Courier New	Programming code: `AutoLISP, VBA` Command line output: `LIST OBJECTS`

Product Benefits and Features

Release 14 offers several new features and benefits over previous versions to increase its usability and functionality. The following sections briefly describe how five key areas of AutoCAD Release 14 benefit the user.

Drawing Productivity

In Release 14, drawing has never been easier. You can draw faster and more accurately with Release 14's advanced performance tools, quick precision drawing tools, and familiar Windows interaction with AutoCAD and your drawing.

Presentation-Quality Drawings

With Release 14, you can polish your drawings for presentation by adding solid fills, graphic art, and high-quality renderings. Also, your drawings can be further perfected using the improved OLE automation feature.

Shared Designs

Release 14 makes it easier to reuse design information and share the information within your project team. This is accomplished by supporting clipped external reference files, embedding raster images, and providing useful Internet tools.

Customization and Support

You can integrate Release 14 with other applications by using ActiveX Automation. Also, you can save time using persistent AutoLISP and application demand loading, and improve object interoperability with ObjectARX.

Management Tools

You can simplify installation, configuration, and management of your AutoCAD network using various management tools. Such tools include the Network Installation Wizard, user profiles, network printing, file sharing, and use of the Autodesk License Manager.

Chapter 1

Introduction to AutoCAD R14

The purpose of this chapter is to learn how to navigate the AutoCAD user interface, which adheres to the Microsoft Windows standard user interface, and makes AutoCAD an easy-to-learn, productivity enhancing CAD tool. Some of the user interface features covered in this chapter include: menu layout and organization, toolbar control, tooltips, scroll bar, and window controls.

About This Chapter

In this chapter, you will do the following:

▶ Start AutoCAD from the Windows Start menu.

▶ Explore the various Help functions of the Help menu.

▶ Explore and practice using menus, dialog boxes, and toolbars.

▶ Explore the Preferences dialog box.

▶ Work with the command window.

Platforms and System Hardware Requirements

This section identifies the operating systems for which Autodesk provides Release 14 support and the minimum and recommended requirements for running Release 14.

Platform Support

AutoCAD Release 14 supports three major operating systems: Windows 95, Windows NT 3.51, and Windows NT 4.0.

System Requirements

The following list consists of recommended hardware and software requirements for proper operation of AutoCAD Release 14:

- Windows NT 3.51 or 4.0 or Windows 95

- 32 MB of RAM

- 50 MB of hard disk space

- 64 MB of disk swap space

- 10 MB of additional RAM for each concurrent AutoCAD session

- 3.5 MB of free disk space during installation only (this space is used for temporary files that are removed when installation is complete)

- With Windows NT 3.51 only: Service Pack 4 or 5 if you install Internet Utilities with the custom or full installation method

- Intel 486, Pentium 90, or better processor

- 1024 by 768 SVGA display

- 4X speed or faster CD-ROM drive for initial installation

- For international single-user and educational versions only: IBM-compatible parallel port and hardware lock

- Mouse or other pointing device

Starting AutoCAD

In the process of installation, the AutoCAD program group is placed in the *c:/program files/autocad r14* directory by default.

To start AutoCAD, choose the Windows Start button on your screen. Choose Programs, AutoCAD R14, then choose AutoCAD R14, as shown in the following figure:

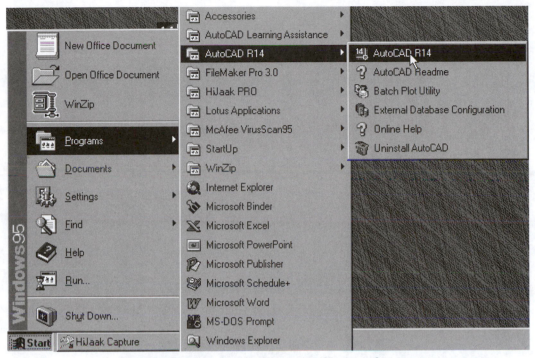

Figure 1-1: Starting AutoCAD from the Windows Start menu

Also in the process of installation, AutoCAD places a shortcut icon on your Windows desktop. Double-click this icon to quickly access AutoCAD.

Using the Help Menu

Before you start an in-depth exploration of AutoCAD, note that AutoCAD provides you with a variety of learning tools from which to choose. These include extensive online help and tutorials. The starting point for these learning tools is the AutoCAD Help dialog box which can be accessed by the following methods:

▶ Toolbar: Standard

▶ Menu: Help > AutoCAD Help Topics
Command: Help or ?
Function Key: F1

The AutoCAD Help dialog box is displayed, as shown in the following figure:

Figure 1-2: The AutoCAD Help dialog box

The AutoCAD Help dialog box is useful for finding information about any of the various standard AutoCAD commands, system variables, and point entry formats. At any time during an AutoCAD session, you can access the online Help menu by pressing the F1 function key.

Each of the following three Help tabs organizes information in slightly different ways:

- *Contents* - displays general topics by category name

- *Index* - displays an alphabetical listing of all search keywords, similar to an index you would find in the back of a printed manual. With each letter you enter, the listing changes to reflect the possible topics related to your entry.

- *Find* - Creates a word list based on all keywords in all the online help files. You can refine this list using the Options button in the Find area. This option should be used as a last resort.

There are additional features that assist you in gaining a better understanding of the AutoCAD program capabilities. You access these help options from the AutoCAD Help Menu:

- *AutoCAD Quick Tour* - For new users. A multimedia overview of basic AutoCAD concepts and capabilities

▶ *AutoCAD What's New in Release 14* - For users upgrading from Release 13, multimedia presentation highlighting the significant new features found in AutoCAD Release 14

▶ *AutoCAD Learning Assistance* - The interactive multimedia training tool that reviews more than 150 specific drawing techniques in 50 lessons. These lessons were designed to teach new and experienced users specific skills that will make them more productive. Each lesson takes approximately five minutes to review, and there is a Try It button which lets you follow along with each lesson, if desired.

▶ *Connect to Internet* - Connects you to the Autodesk web site (*www.autodesk.com/autocad*) where you can find information on Autodesk products, support and training, company information, and industry solutions.

▶ *About AutoCAD* - Displays the AutoCAD version number, serial number, license information, dealer name, and copyright information.

Using Menus, Dialog Boxes, and Toolbars

Pull Down Menus

When you perform a Full installation of AutoCAD, you are presented with a menu bar containing 11 menus: File, Edit, View, Insert, Format, Tools, Draw, Dimension, Modify, Bonus, and Help. These menus provide an alternate method of accessing commands and dialog boxes other than the Command prompt or a toolbar button. They are termed *pull-down menus* because when you choose one with your left mouse button, the menu is displayed beneath its title, as if you were pulling the menu down from its title. The menu remains present until you click anywhere outside of it. Click a menu option to choose it. You will notice that some menu items are shaded and cannot be chosen at certain times. This means that a particular action is unavailable at that point in your drawing session. For example, in the following figure, the Paste option is shaded. This is because nothing has been copied to the clipboard for you to paste into your drawing.

Note: Performing a Typical installation of AutoCAD provides you with 10 menus. The Bonus menu is not included in a Typical installation.

The menu bar and the default Edit menu are shown in the following figure:

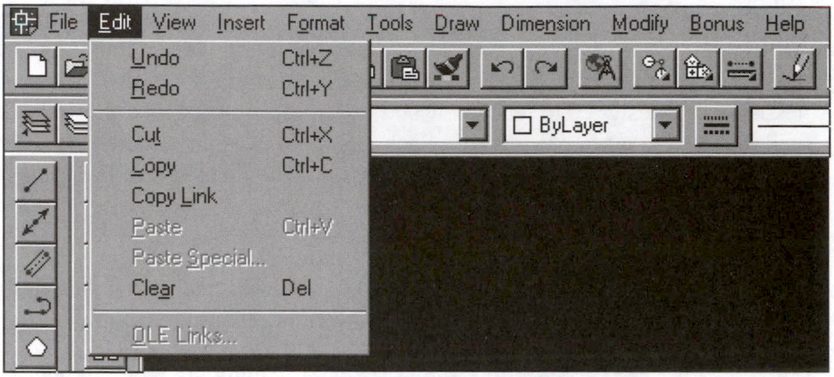

Figure 1-3: The menu bar with the expanded Edit menu

Cascading Menus

Several menu options lead you to other related menus, termed *submenus*, which contain further command options. These types of menus are termed *cascading*, or *hierarchical* menus. Menu options which lead to submenus display a small black arrowhead to the right of the option, as shown in the following figure. Moving your cursor over the option automatically displays the submenu.

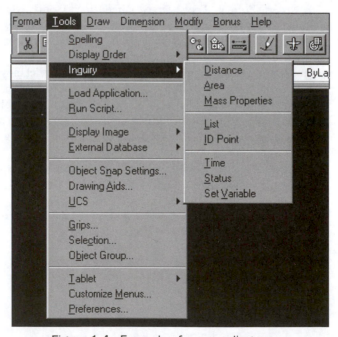

Figure 1-4: Example of a cascading menu

Cursor-Menu Default Options

The cursor menu works like the menu bar menus, except that it is not a permanent part of the application window. You can access the default cursor menu which provides quick access to frequently used menu items. To do so press the SHIFT key while depressing the right mouse button, or depress the right mouse button while holding the SHIFT key down.

The default cursor menu is shown in the following figure:

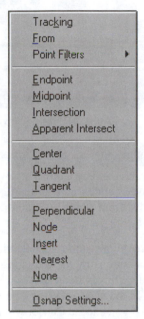

Figure 1-5: The default cursor menu

Dialog Boxes

If you choose a menu option with trailing ellipses (...) you will open one of the many AutoCAD dialog boxes. Dialog boxes provide a visual, intuitive interface for executing various commands. A typical dialog box is shown in the following figure:

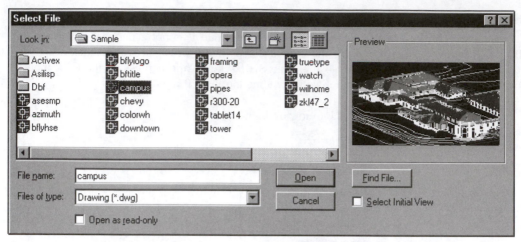

Figure 1-6: A typical AutoCAD dialog box

You will find many of the following features in most dialog boxes:

▶ *Drop-down lists* - Characterized by a down arrow to the right of a field. By default, they are displayed condensed with the first option visible. Choosing the arrow expands the list from which you can then select any available option.

▶ *Buttons* - Appear as three-dimensional shapes raised above the surface of the dialog box. Choosing a button always results in some immediate action. The three most common dialog box buttons are the OK button, the Cancel button, and the Apply button. If you choose OK or Apply, you cause your dialog box settings to take effect. If you choose Cancel, your dialog box setting will not take effect.

▶ *Fields* - Look the same as a drop-down list without the arrow. In fields, you enter the name of an option or drawing. You are not given a list of items from which to choose.

▶ *Checkboxes* - Small boxes with various options to the right of them. Choosing the checkbox will place a check in the box enabling the related option. This action will take effect once you choose the OK or Apply button.

▶ *Radio buttons* - Small circles associated with options. These are similar to checkboxes except that in a set of radio button options, only one can be selected, turning the others off. With checkboxes, one, none, or all can be checked and applied at the same time.

▶ *Image tiles* - Display a small representation of how a feature will look in a drawing, or how a drawing will look full size.

▶ *Scroll bars* - Work the same way as the drawing window scroll bars. Dialog box scroll bars are present when there are too many files or options in a list to view at once.

◗ *Tabs* - Sections or "pages" of a dialog box containing expanded information that relates to the general topic of the dialog box

◗ List boxes - By default, scrollable list boxes of items you can select. They are not collapsible.

To close a dialog box, choose OK or Cancel, or choose the "X" button in the top right corner of the title bar.

Toolbars

Another method of performing commands is through the various toolbar tools. There are many tools in the 21 toolbars, but those most frequently used, relating to file, edit, and view, are found in the Standard toolbar, shown in the following figure:

Figure 1-7: The Standard toolbar

To display the other toolbars, you can access the Toolbars dialog box in the following ways:

◗ **Toolbar**: Right-clicking any currently visible toolbar

◗ **Menu:** View > Toolbars

◗ **Command:** TOOLBAR

The Toolbars dialog box is displayed, as shown in the following figure:

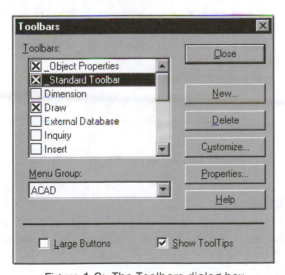

Figure 1-8: The Toolbars dialog box

Checking any of the toolbar checkboxes displays the toolbar. The toolbar can then be moved anywhere in the application window by selecting the title bar and dragging the toolbar to a location.

Buttons

Button is the name commonly used for a tool. A toolbar button is identified by its icon, or image, which represents the command that the tool accesses. For example, the SAVE command has a tool button located in the Standard toolbar with an icon of a floppy disk, as follows:

When you move your cursor over a button, you see a *tooltip* which shows the name of the command. Tooltips help you identify commands, as shown in the following figure:

Figure 1-9: Example of a tooltip

Flyouts

Flyouts are extensions of certain toolbar buttons which provide related tools or command options while keeping toolbars compact. The buttons containing flyouts have a small black arrow at the bottom right corner of the button. A single click on the button executes the top command. If you hold down the left mouse button on the button containing a flyout, the flyout will be displayed. You can then move the cursor over each button, reading the tooltips to determine which option you want. When you locate the correct button, place your cursor over it and release the left mouse button. This option becomes the top or default option for subsequent use.

Setting Preferences

Using the Preferences dialog box, you can change many aspects of your drawing window according to personal preferences. In this section, you learn about changing the background color, changing the size of the cursor, and disabling scroll bars.

Preferences Dialog Box

Methods of invoking the Preferences dialog box include:

▶ **Menu:** Tools > Preferences

▶ **Command:** PREFERENCES

The Preferences dialog box is displayed, as shown in the following figure:

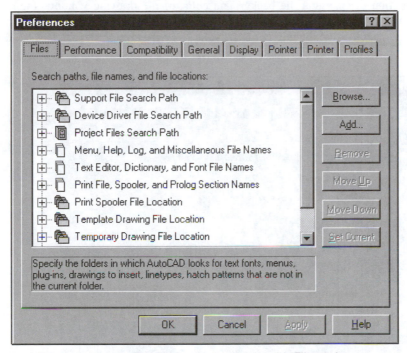

Figure 1-10: The Preferences dialog box with Files tab open

The Preferences dialog box contains the following eight tabs which organize and compile the preferences settings and system configuration:

▶ *Files* - Stores specific search paths, filenames, and file locations. Three icons represent different structures for locating individual files and special folders. The folder icon is for search paths, the paper stack icon is for specific files, and the file cabinet icon is for a specific folder.

▶ *Performance* - Controls performance enhancing system variables.

▶ *Compatibility* - Sets options for users upgrading to Release 14 from previous releases.

▶ *General* - Controls system variables for operating parameters.

▶ *Display* - Controls settings for the drawing and text windows.

▶ *Pointer* - Sets the current pointing device.

▶ *Printer* - Adds and configures printers.

▶ *Profiles* - Saves and restores user preference settings.

Color

You have the option of changing the background color of the drawing window, the AutoCAD Text Window, and the command window to any shade and hue of 16 colors. You can also use the sliders to set the colors to a combination of RGB values. To do this, access the Display tab of the Preferences dialog box, shown in the following figure:

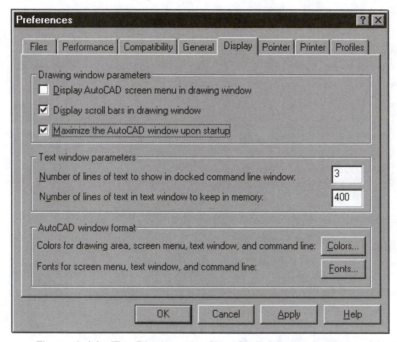

Figure 1-11: The Display tab of the Preferences dialog box

Choosing the Colors button in the AutoCAD Window Format area opens the AutoCAD Window Colors dialog box, as shown in the following figure:

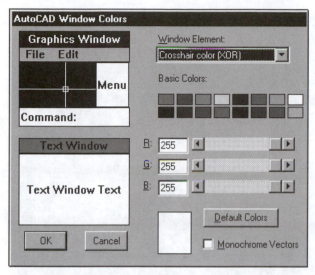

Figure 1-12: The AutoCAD Window Colors dialog box

Choose the option you want to change in the Window Element drop-down list, select a basic color and change its shade or hue. Choose OK to close both dialog boxes and your changes will take effect. Instead of choosing OK in the Preferences dialog box, you can also choose Apply, which applies the change but does not close the dialog box. You can then go to other tabs and make other changes before closing the dialog box.

Cursor

You can change the size of the drawing cursor, making it a certain percentage of the screen size, using the Pointer tab of the Preferences dialog box, as shown in the following figure:

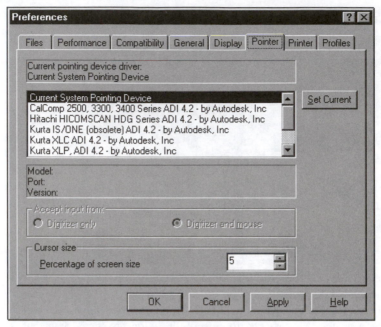

Figure 1-13: Pointer tab of the Preferences dialog box

In the Cursor Size area, you can set your desired cursor size by scrolling up or down using the spinner arrows, or by entering a percentage from 1 to 100 in the Cursor Size field.

Scroll Bars

The vertical and horizontal scroll bars in the drawing window provide you with an infinite range of movement in moving the drawing view horizontally and vertically. Click the up and down, right and left arrows once to move one increment at a time, or press and hold the left mouse button down to move a greater distance. You can also use the thumbwheel, shown in the following figure, to scroll more smoothly but less precisely through the view. Drag the thumbwheel along the scroll bar.

Figure 1-14: Horizontal scroll bar with thumbwheel

If you want to add greater space to your drawing area, you can hide the vertical and horizontal scroll bars. To do this, access the Display tab of the Preferences dialog box. In the Drawing Window Parameters area, clear the Display Scroll Bars in Drawing Window checkbox, which is checked by default.

The Command Window

The command window is a docked or floating window where you can enter names of commands and where AutoCAD displays prompts and messages.

By default, the size of the command window is equivalent to three lines of text and is the width of the application window. It is also located by default at the bottom of the application window. However, the command window can be vertically resized using the splitter bar, which the cursor becomes when placed at the top edge of the window. Dragging the splitter bar up or down resizes the window to your required height. In the process of enlarging the command window, you are displaying the command history.

The default command window is displayed, as shown in the following figure:

Figure 1-15: The command window

Command Aliases

Once you become comfortable entering commands at the Command prompt, you can learn the command aliases, which are basic one-, two-, or three-letter abbreviations of the commands. For example, you can enter **X** at the Command prompt to access the EXPLODE command rather than entering **explode** at the Command prompt. Many commands have an alias, and learning these can help you work faster and more productively. A list of all the command aliases are located in the Help menu.

The Command Prompt

When you enter a command in the command window, AutoCAD either displays a dialog box or prompts you for further information. This line of command text is termed the *Command prompt*. The Command prompt asks you to specify coordinate values, command options, or any other data needed to complete the command. The following text is an example of the Command prompt for the PLINE command:

Arc/Close/Halfwidth/Length/Undo/Width/<Endpoint of line>:

The Default and Uppercase Options

The default option in the Command prompt is the one enclosed in brackets, for example, <Endpoint of line>. To choose this option, press the ENTER key. To choose any of the other options available at the Command prompt, enter the uppercase letter(s) of the options. Command options can be entered with upper or lowercase text. Then press the ENTER key. For example, in the Command option prompt for the PLINE command:

Arc/Close/Halfwidth/Length/Undo/Width/<Endpoint of line>:

entering **A, a** or **arc** will create an arc polyline.

Transparent Commands

You can use many commands transparently. This means that you can enter certain commands at the Command prompt while you are using another command. Transparent commands are not used to select objects, create new objects, end drawing sessions, or cause regenerations. Changes that you make in dialog boxes that have been opened transparently cannot take effect until the interrupted command is executed. Also, when you reset a system variable transparently, the new value does not take effect until you start the next command.

To use a transparent command while in a current command, enter an apostrophe (') before entering a transparent command at the Command prompt. Transparent commands are also selected by choosing the command from the AutoCAD menu bar or choosing a valid transparent command button from a toolbar. Double angle brackets precede prompts for transparent commands. After you complete the transparent command, the original command resumes. For example, you may want to zoom in on a portion of a drawing while you are drawing a line. The Command prompt information may look like the following:

Command: line

From point: 'zoom

>>All/Center/Dynamic/Extents/Previous/Scale(X/XP)/Window/<Realtime>: w

>>First corner: >>Other corner:

Resuming LINE command.

From point:

To point:

To point:

You can access a list of all the transparent commands through the Help menu.

Editing Entered Commands

One extremely helpful feature of AutoCAD is the ability to edit commands once you enter them at the Command prompt. For example, if you are trying to draw a line and you incorrectly enter "lene" at the Command prompt, you can use the left and right keyboard arrow keys and the DELETE key to edit the word. You can also use the BACKSPACE key and the Insert mode.

Pressing ENTER to Repeat the Last Command

Another convenient feature of the AutoCAD Command prompt is the ability to repeat the last command you entered. Once you have completed or canceled a command, press the ENTER key once to recall and execute the last command.

Recalling Previously Entered Commands

AutoCAD stores in memory each command you enter at the Command prompt in a given session. You can recall previous commands by pressing the "up" keyboard arrow. If you press ENTER when the desired command is displayed at the Command prompt, AutoCAD will execute the command.

Pressing ESC to Cancel

At any point while you are performing commands at the Command prompt, or from a menu or toolbar, you can cancel the command completely by pressing the ESC key on your keyboard.

Online Help and Documentation

Online Help and documentation, accessed from the Help menu, is an updated, enhanced version of the Release 13 documentation. This Online Help is presented in three forms, summarized as follows:

▶ *AutoCAD Help Topics* - The new AutoCAD Help dialog box contains the Contents, Index, and Find tabs, as shown in the following figure. You can also access the AutoCAD Help dialog box at the Command line. If you are unsure of the correct Command line syntax, or if you are in the middle of performing a command and you are unsure of the next step, press F1 to open the AutoCAD Help dialog box.

▶ *What's New* - This Help menu option provides you with an interactive list of new AutoCAD Release 14 features. When you select What's New from the Help menu, a screen is displayed listing the new feature topics. Placing your cursor on a topic button displays a text box containing a brief description of the AutoCAD improvement. Choosing a topic button displays a screen that lists the new features relating to the chosen topic. When you choose a feature, an animated screen is displayed containing text boxes which demonstrate how the feature works.

▶ *Quick Tour* - The main menu of this Help option contains two buttons: Introducing AutoCAD Release 14, and Drawing with AutoCAD. When you choose a button, a list of functional topics is displayed, such as Drawing a Room in a Floor Plan, and Accessing Online Documentation.

Choosing a topic displays an animated screen containing text boxes that quickly guide you through an example of how to perform a particular function in Release 14.

Figure 1-16: The AutoCAD Help dialog box

The Contents Tab

The Contents tab displays the names of the online documentation. An example of how these documents are linked together is clearly illustrated in the How To section of this tab.

To display a typical step-by-step tutorial, as shown in the following figure, choose How to, then choose Set Up a Drawing, then choose Set the Plot Scale.

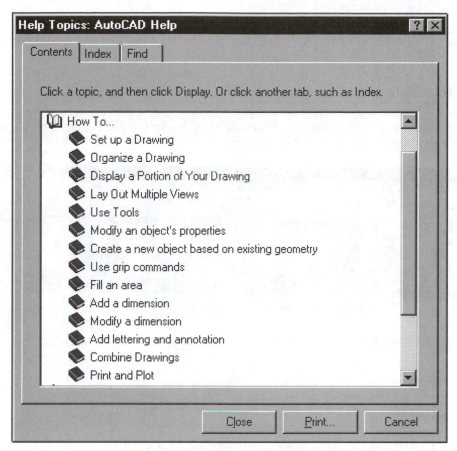

Figure 1-17: A typical How To Help screen

In addition to the tutorial, the relevant section in the *User's Guide* can be accessed by choosing Show User's Guide, which is located near the top of the screen.

The Index Tab

The Index tab of the AutoCAD Help dialog box displays an alphabetical listing of all search keywords, synonymous with an index you would find in the back of a printed manual. With each letter you enter, the listing changes to reflect the possible topics related to your entry. For example, entering the letter **F** brings up all the possible topics that begin with the letter **F**; entering **Fi** narrows the search to items beginning with Fi , and so on. Once you enter your full search word, choose from the list of possible topics. For example, if you enter **line**, double-click on the Line possibility in the list box to open a second dialog box with a more detailed listing of the topics that relate to your selected word. Double-click on one of those possibilities to access specific topical information.

The Find Tab

The Find tab of the AutoCAD Help dialog box creates a word list based on all keywords in all the online help files. You can refine this list using the Options button in the Find area.

"What's This" Help

A new help feature in Release 14 is the What's This option. Where a ToolTip does not exist, you can place the cursor over an item in a dialog box and press the right mouse button. This displays the What's This option as shown the following figure:

Figure 1-18: The What's This display

Pressing the left mouse button on What's This displays a help screen as shown in the following figure:

> Lists the available folders and files. To see how the current folder fits in the hierarchy on your computer, click the down arrow. To see what's inside a folder, click it.
>
> The box below shows the folders and files in the selected location. You can also double-click a folder or file in that box to open it.
>
> To open the folder one level higher, click 🔼 on the toolbar.

Figure 1-19: Example of What's This help screen

Printed Guides

A reference card and two printed manuals are provided with AutoCAD Release 14 to facilitate your learning.

- ▶ The Quick Reference Card, a helpful tool for all users of AutoCAD, is located at the back of the *User's Guide*. This card provides an "at-a-glance" reference to toolbars, object selection methods, keyboard shortcuts, and coordinate entry methods. This card is not available online.

- ▶ The *Installation Guide* contains the same material as the online Installation Guide, except in printed format.

- ▶ The *User's Guide* contains the same material as the online User's Guide, except in printed format.

You can also order the printed Command Reference and Customization Guide as a set by ordering the AutoCAD Release 14 Documentation Pack through your authorized Autodesk dealer or reseller.

AutoCAD Learning Assistance

The AutoCAD Learning Assistance (ALA) is a new interactive, multimedia training tool that provides an online learning environment in which you can quickly gain the skills needed to increase productivity during the software learning curve. With ALA, you can master techniques for creating, editing, and maintaining drawings in a production environment.

Figure 1-20: Multimedia Learning Assistance

AutoCAD Learning Assistance offers more than 50 lessons covering 160 skills, including how to:

- Create template files

- Adjust dimensions for multiscale drawings

- Place Microsoft Excel tables in drawing files with Object Linking and Embedding (OLE)

- Use enhanced imaging tools to scale, clip, and insert aerial maps and other raster-based objects into vector-based AutoCAD drawings

◗ Create toolbars to automate common tasks

◗ Link drawing data to Web pages using built-in Internet utilities

Most ALA exercises can be reviewed in less than five minutes and provide an immediate productivity boost. By following ALA's "watch-and-try" technique, you quickly learn to adopt and adapt new drawing methods. You can review lessons in a predetermined sequence by way of a user-driven course, or select only those lessons relevant to a particular project on which you are working.

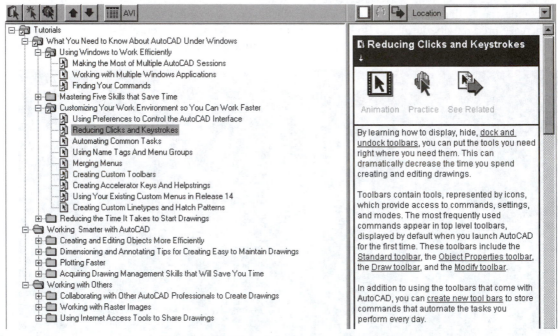

Figure 1-21: Tutorials screen section

New or experienced users who need more information about specific functions or tools can follow dynamic links to access the online reference materials. Also, the *Fast Answers* interface helps you resolve technical problems as you work. For example, designers who are having trouble with the PLOT command or with the dimension styles can immediately access the relevant lesson. You can follow topics and commands from lesson to lesson, command to command.

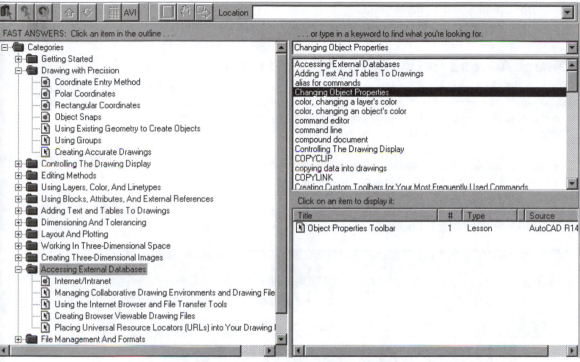

Figure 1-22: Fast Answer screen section

The *Concepts* section provides you with animated overviews of concepts that will help you work more efficiently with Release 14. When you choose the Concept button, an alphabetical list of topics is displayed. When you choose an item, a screen is displayed with information about the topic. You can also learn more about your topic by choosing the Animation icon or the See Related icon, which allow you to view a list of related topics.

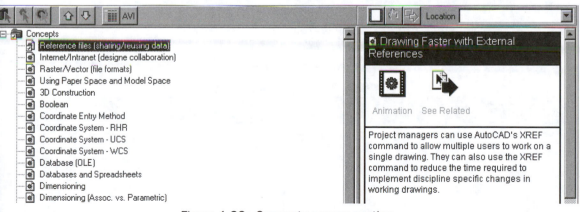

Figure 1-23: Concepts screen section

AutoCAD Learning Assistance also links you directly to the Autodesk Web site, where you can find information about in-depth training, telephone support, and on-sight assistance available through the worldwide network of Autodesk Training and Systems Centers.

Direct Access to the AutoCAD Web Page

Many AutoCAD users have been able to get online support and download the latest information on products, drivers, and incremental releases by accessing the Autodesk Web page. Located at *www.autodesk.com/autocad,* the page contains up-to-the-minute information on Autodesk. Options in accessing the Autodesk Web page include:

 ❿ Toolbar: Standard

 ❿ Menu: Help > Connect to Internet

 ❿ Command: BROWSER

The Autodesk Web page is displayed, as shown in the following figure:

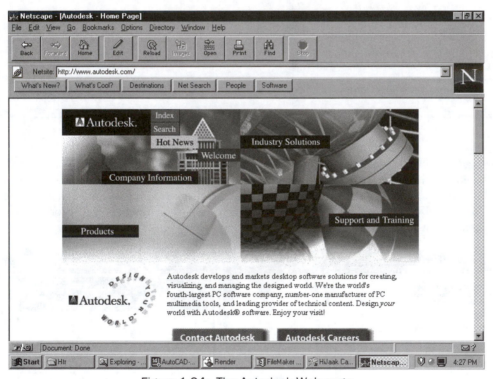

Figure 1-24: The Autodesk Web page

Exercise 1-1: Working with the User Interface

In this exercise, you start AutoCAD from the AutoCAD R14 program group. Then you set options in the Toolbars dialog box and in the Preferences dialog box. Finally, you view previously entered commands with the AutoCAD text window.

Starting AutoCAD

1. From your Start menu, choose Programs, then choose AutoCAD R14, then choose AutoCAD R14. Your AutoCAD application window is displayed containing the Start Up dialog box.

2. In the Start Up dialog box, choose the Start from Scratch button, then choose OK. A default drawing is displayed.

Using the Toolbars Dialog Box

1. At the Command prompt, enter **toolbar**. The Toolbars dialog box is displayed, as shown in the following figure:

Figure 1-25: The Toolbars dialog box

2. In the Toolbars list box, check the Modify II checkbox. The Modify II toolbar is displayed in the drawing window as a floating toolbar.

3. Now you will change the Modify II toolbar from a floating toolbar to a docked toolbar. To do this, place the cursor on the title bar of the toolbar, then press and hold the left mouse button.

4. Drag the Modify II toolbar to the docking region under the Object Properties toolbar, then release the left mouse button. When the toolbar is docked, the title bar name is no

longer displayed. The docked toolbar is shown in the following figure:

Figure 1-26: The docked Modify II toolbar

5. In the Toolbar dialog box, clear the Modify II checkbox. The Modify II toolbar is no longer displayed.

6. Choose Close to close the Toolbars dialog box.

Setting Options in the Preferences Dialog Box

1. From the Tools menu, choose Preferences. The Preferences dialog box is displayed.

2. If the Display tab is not the current tab, choose the Display tab to make it current. The Display tab of the Preferences dialog box is displayed, as shown in the following figure:

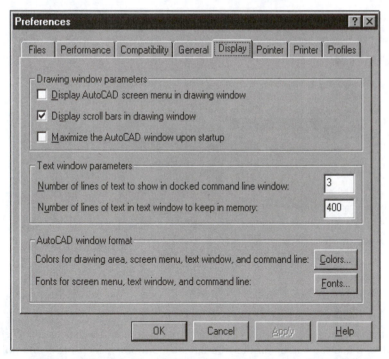

Figure 1-27: The Display tab of the Preferences dialog box

3. In the Drawing Window Parameters area, check the Maximize the AutoCAD Window Upon Startup checkbox.

4. Choose OK to apply the changes and close the Preferences dialog box.

5. Press ENTER to repeat the last command which opens the Preferences dialog box again.

6. Clear the Display Scroll Bars in Drawing Window checkbox, then choose the Apply button. The scroll bars are no longer displayed in the drawing window.

7. Choose the OK button to close the Preferences dialog box.

Using Cascading Menus

1. Some AutoCAD commands are accessed from cascading menus. These commands are identified by a triangular symbol located next to the name in the pull-down menu.

   Choose Tools from the AutoCAD menu bar. The Tools pull-down menu is displayed, as shown in the following figure:

Figure 1-28: The Tools pull-down menu

2. Postion the cursor over Inquiry. The Tools menu with the Inquiry cascading menu is displayed, as shown in the following figure:

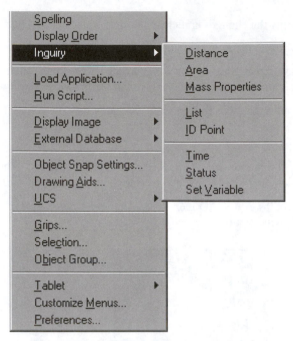

Figure 1-29: The Inquiry cascading menu

3. Now choose Status. The AutoCAD text window is displayed, as shown in the following figure:

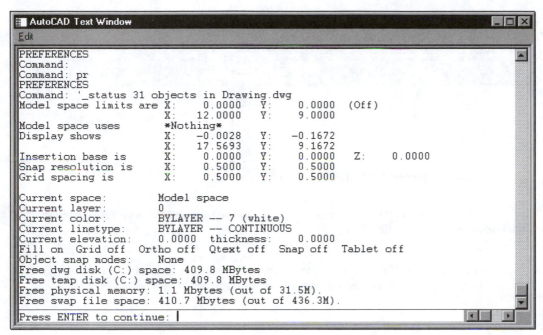

Figure 1-30: The AutoCAD Text Window

4.   Press ENTER to continue and then press F2 to close the AutoCAD Text Window.

5.   You can also start the STATUS command from the command prompt, enter **status**. After the AutoCAD Text Window is displayed, press the ESC key to cancel the command. Press the F2 key.

6.   Press the up or down arrow keys to cycle through the previously-entered commands. Commands entered through the CLI are displayed in the command window.

Using Toolbar Flyouts

1.   Some toolbar buttons have a small black triangle located in the lower-right corner of the button. This triangle indicates that a toolbar with associated buttons is displayed when the flyout button is selected. The Distance button with flyout symbol is shown in the following figure:

Figure 1-31: The Distance button displaying the flyout symbol

2.   In the Standard toolbar, choose the Distance button by pressing and hold down the left mouse button until the Inquiry toolbar flyout buttons are displayed, as shown in the following figure:

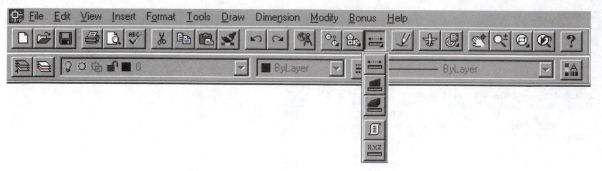

Figure 1-32: The Inquiry toolbar flyout

3.   Continue holding down the left mouse button. Position the cursor over the Locate Point button located at the bottom of the flyout as displayed in figure 1-27. The Locate Point tooltip is displayed.

4.   Choose the Locate Point button. Position the cursor over any area in the drawing window and press the left mouse button. The coordinates of the point you selected are displayed in the CLI.

5.   Notice that the Locate Point button has replaced the Distance button as the first one in the Inquiry toolbar flyout.

6.   Press ESC to cancel the command.

Conclusion

After completing this chapter, you have learned the following:

▶   You have learned to start AutoCAD from the Start menu by choosing Programs, then choosing AutoCAD R14 program group, then choosing AutoCAD R14.

▶   The command window is docked at the bottom of the application window and is the place where you enter commands and view prompts and messages.

▶   The keyboard function keys are useful as shortcuts to commands, modifiers, and the Help menu.

▶   The AutoCAD Help dialog box is useful for finding information about any of the standard AutoCAD commands, system variables, and point entry formats.

▶   You can view hidden toolbars by checking various checkboxes in the Toolbars dialog box.

▶   You can change the background color of your drawing window and resize the cursor using the Display tab of the Preferences dialog box.

Chapter 2

Working with Drawings

In this chapter, you learn how to save new drawings and existing drawings, control drawing display, create and modify line segments, and select and remove objects in a drawing.

About this Chapter:

In this chapter, you will do the following:

▶ Use the NEW command to start a new drawing.

▶ Use the OPEN command to display existing AutoCAD drawings.

▶ Use the SAVE command to save a current drawing.

▶ Use the PAN and ZOOM commands to change the drawing display view area and magnification.

▶ Create line segments using the LINE command.

▶ Modify line segments using the LINE command Close and Undo options.

▶ Draw circles and arcs using the CIRCLE and ARC commands.

Starting a New Drawing

Each AutoCAD drawing session begins with a default drawing named *drawing.dwg*. You use this default file to create new drawings based on settings saved in the default drawing.

Methods for invoking the NEW command include:

▶ **Toolbar**: Standard

▶ **Menu:** File > New

▶ **Command**: NEW

Use the following steps to create a new drawing:

1. At the Command prompt, enter **new**.

2. The Create New Drawing dialog box is displayed. The Create New Drawing dialog box is shown in the following figure.

3. In the Create New Drawing dialog box, choose the Start from Scratch button.

4. In the Select Default Setting list box, select English or Metric, then choose OK.

5. When you are ready to assign a filename to your drawing, choose Save or Save As from the File menu. The Save Drawing As dialog box is displayed.

6. In the Save Drawing As dialog box, in the File Name field, enter a name for the drawing and choose OK. To view the Save Drawing As dialog box see Figure 2-5.

The Create New Drawing dialog box is displayed, as shown in the following figure:

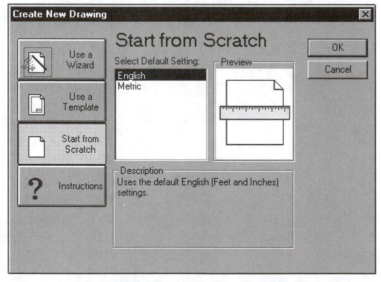

Figure 2-1: The Create New Drawing dialog box

Opening an Existing Drawing

You can access existing drawing files by using the OPEN command. Drawings are primarily opened using the Select File dialog box, but they can also be opened from the Browse/Search dialog box.

Methods for invoking the OPEN command include:

> ◗ **Toolbar:** Standard

> ◗ **Menu:** File > Open

> ◗ **Command:** OPEN

When you invoke the OPEN command, AutoCAD displays the Select File dialog box, as shown in the following figure. All AutoCAD drawing files with a *.dwg* file extension in the selected folder are displayed. You can also select *.dxf*, template files and files from earlier releases of AutoCAD.

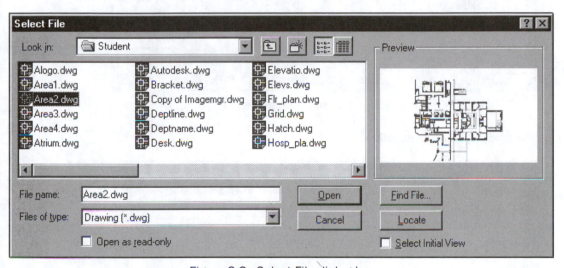

Figure 2-2: Select File dialog box

The Select File dialog box has a Preview image tile that displays a bitmap of the selected file if a bitmap image was generated with Release 13 or Release 14.

The Find File option searches multiple paths and drives using search criteria. When you choose the Find File button, the Browse/Search dialog box is displayed. The Browse tab is shown in the following figure:

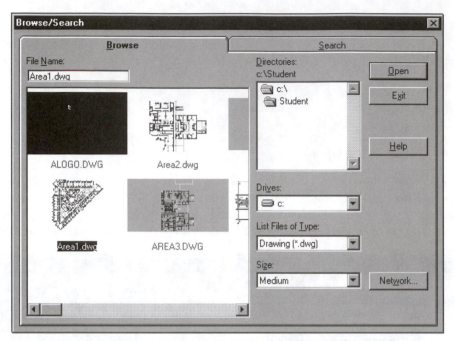

Figure 2-3: The Browse tab of the Browse/Search dialog box

The Search tab is shown in the following figure:

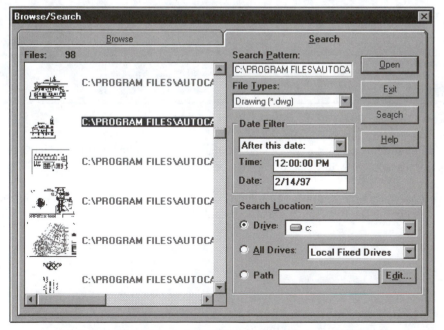

Figure 2-4: The Search tab of the Browse/Search dialog box

Saving the Current Drawing

AutoCAD uses different file saving commands that protect your work by storing the existing drawing status to a named file in a directory. Saving your work is very important and should be done every 10-15 minutes. If there is a power failure, or a systems failure, all of your work saved prior to the problem will be usable. All unsaved work may be lost or irretrievable. Not saving your work regularly can result in hours of lost time and wasted energy.

Various file saving commands are used to store drawing information. These commands include: SAVE, SAVEAS, and QSAVE. The SAVE command saves the drawing with the current filename or a specified name when saved the first time.

> ▶ Command: SAVE

The SAVEAS command saves unnamed drawings with a filename, or saves the current drawing with a different name. Methods for invoking the SAVEAS command include:

> ▶ **Menu:** File > SaveAs

> ▶ **Command:** SAVEAS

The QSAVE command saves currently named drawings one at a time without requesting a filename. Methods for invoking the QSAVE command include:

> ▶ Toolbar: Standard

> ▶ Menu: File > Save

> ▶ Command: QSAVE

Use the following steps to save unnamed drawings:

1. From the File menu, choose Save.

2. If you previously saved and named a drawing, AutoCAD will update the directory with any new information. If the drawing has never been saved, the Save Drawing As dialog box will be displayed.

3. In the Save Drawing As dialog box, in the File Name field, enter the new drawing name (the file extension is not required).

4. Choose OK.

The Save Drawing As dialog box is shown in the following figure:

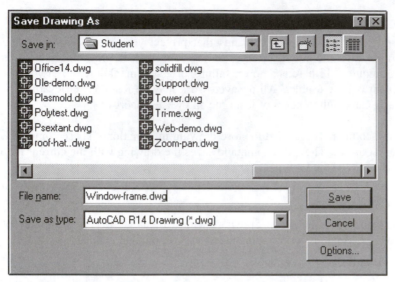

Figure 2-5: The Save Drawing As dialog box

Controlling the Display with PAN/ZOOM

Display commands let you view areas of drawings while they are being developed and after they are completed. All of the display command options effect the way a drawing is viewed. These options include zooming and panning, extents, plan view and 3D views, model space and paper space, viewports, and resolution. In this section we will discuss the PAN and ZOOM display commands.

PAN Command

The PAN command lets you move the drawing display in the current viewport in any direction. The PAN command can be used as a transparent command which means you can use it while inside another command.

The PAN command default is Realtime Pan. When you select the Realtime Pan command option, the cursor changes into a hand cursor. To change the location of your drawing, press the left mouse button. This locks the cursor to its current location relative to the viewport coordinate system. As you move the mouse, the drawing image pans to a new location. Graphics within the drawing window are moved in the same direction as the cursor.

Methods for invoking the Realtime PAN command default include:

- **Toolbar:** Standard

- **Menu:** View > Pan >Real Time

- **Command:** PAN

ZOOM Command

The ZOOM command lets you increase or decrease the magnification of images in the view window. When you zoom out you can see larger portions of a drawing. Zooming in lets you enlarge drawing portions, enabling you to see the image with greater detail. Zooming does not change the absolute size of the drawing. It does, however, change the relative magnification of objects within the drawing window.

Methods for invoking the ZOOM command include:

▶ **Toolbar:** Standard > Zoom Flyout

▶ **Menu:** View > Zoom

▶ **Command:** ZOOM

AutoCAD offers several general ways to change the view using ZOOM. These include specifying a display window, zooming to a specific scale, and displaying the entire drawing. There are also twelve ZOOM command options to assist you in creating specific views: Realtime, In, Out, All, Center, Dynamic, Extents, Previous, Window, Scale (X), and Scale times paper space (XP). In this section we will discuss the ZOOM All, ZOOM Window, ZOOM Extents, and ZOOM Previous command options.

All

ZOOM All lets you view the entire drawing in the current viewport. In a plan view, AutoCAD zooms to the drawing limits or the current extents, whichever is greater. The display shows all objects even if they extend outside the drawing limits. Use the All option after changing the limits if you want to display the drawing limits.

Methods for invoking the ZOOM All command option include:

▶ **Toolbar:** Standard

▶ **Menu:** View > Zoom > All

▶ **Command:** ZOOM All

Window

ZOOM Window lets you specify the area of the drawing you want to view by entering two opposite corner points of a rectangular window. Objects in the window are displayed larger to fill the drawing window. The region specified by the corners you select is centered in the new display. The region specified in the view window may not exactly match the aspect ratio of the viewport being zoomed.

Methods for invoking the ZOOM Window option include:

▶ **Toolbar:** Standard

▶ **Menu:** View > Zoom > Window

▶ **Command:** ZOOM Window

Use the following steps to specify drawing display boundaries:

1. From the View menu, choose Zoom then choose Window.

2. Select one corner of the area you want to view.

3. Specify the opposite corner of the area you want to view.

The drawing window now displays the new view.

Extents

The Zoom Extents option displays the region of the drawing plane where all objects you draw are located. The display is based just on drawing objects, the drawing limits are not considered to recalculate the display.

Methods for invoking the Zoom Extents option include:

▶ **Toolbar:** Standard > Flyout, Zoom

▶ **Menu:** View > Zoom > Extents

▶ **Command:** ZOOM > Extents

Previous

The Zoom Previous option displays the last view of your drawing. This option lets you restore as many as 10 previous views.

Methods for invoking the Zoom Previous option include:

▶ **Toolbar:** Standard

▶ **Menu:** View > Zoom > Previous

▶ **Command:** ZOOM > Previous

Creating Simple Drawings Using Lines

The LINE command draws a line segment and continues to prompt for points, enabling you to draw continuous lines. The endpoint location of each line segment is specified by two dimensional (X,Y) coordinates or three-dimensional (X,Y,Z) coordinates.

Methods for invoking the LINE command include:

▶ Toolbar: Draw

▶ Menu: Draw > Line

▶ Command: LINE

When you draw line segments you will notice that a "rubberband" line is attached to the last selected point. This line has crosshairs at the endpoint and lets you see exactly where your next line segment will be located.

You can continue to draw line segments until you press the ENTER key or spacebar to exit the LINE command. Each line segment that is drawn is considered a separate object.

CLOSE

The Close option connects the end point of the last line segment to the end point of the first line segment. After you draw a series of two or more continuous line segments, enter **c** or **close** at the `To Point` Command prompt. This ensures that the first and last segments meet precisely, forming a closed loop of line segments.

UNDO

To undo the previous line segment while operating in the LINE command sequence, enter **u** or **undo** at the `To Point:` Command prompt. AutoCAD uses the Undo option to remove line segments in the order they were created in the current LINE command. You then have the option of creating more line segments from that point or ending the LINE command by pressing ENTER.

Line to Line Continuation

The LINE command Continue option lets you draw a new line segment from the endpoint of the most recently drawn line segment. After starting the LINE command, to use this option press ENTER or the spacebar when the From Point Command prompt is displayed.

Exercise 2-1: Drawing Lines with Various Options

 You can use a variety of commands to draw line segments. In this exercise, you use the LINE command and options to draw and modify a series of line segments. You also use display commands to change the view of your drawing.

Note: Most of the exercises contained in this book use files from the CD-ROM. To install these files on your hard drive, run the setup.exe program on the AutoCAD R14 Fundamentals CD-ROM that is packaged with this book.

Creating Line Segments

1.   Start the AutoCAD program. From the File menu, choose Open. The Select File dialog box is displayed.

2.   In the File Name area of the Select File dialog box, enter **line**, then select the Open button.

The drawing looks like the following figure:

Figure 2-6: The initial view of line.dwg

3.   From the Draw menu, choose Line.

4.   Use the cursor left mouse button to select crosshair circle symbol located at point 1, then select the symbol at point 2.

5.   To end the LINE command, press the right mouse button.. The drawing looks like the following figure:

Figure 2-7: Line segment 1-2

Using Pan and Zoom

1. From the View menu, choose Pan, then choose Point.

2. Select a point in the top right corner of the drawing window, then select any point in the center of the drawing window.

3. From the Draw menu, choose Line. Draw a line segment from the symbol located at point 2 to 3, then from point 3 to 4. Press the right mouse button to end the LINE command.

4. From the View menu, choose Zoom, then choose All. The drawing looks like the following figure:

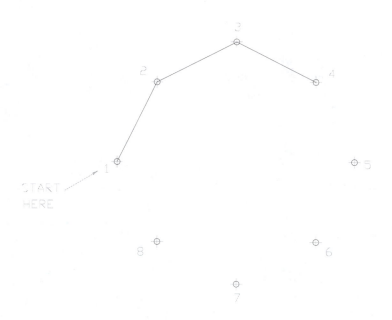

Figure 2-8: Line segments 1-2, 2-3, 3-4

5. Enter **line** at the Command prompt, then press ENTER. Draw a line segment from 4-5; 5-6, 6-7, 7-8, and 8-1, then press the right mouse button to end the line command.

6. Press ENTER to start the LINE command again, then draw a line segment from 6-8 and from 8-2.

7. The line segment that was just drawn was placed incorrectly. Use the Line command option Undo to remove the last line segment. Enter **undo** at the Command prompt, then press ENTER to remove line segment 8-2.

8. Now continue drawing line segments by selecting the point at 8-3.

9. Use the LINE command option Close to complete this series of line segments. Enter **c** at the Command prompt and press ENTER to close the continuous line segment

The completed drawing looks like the following figure:

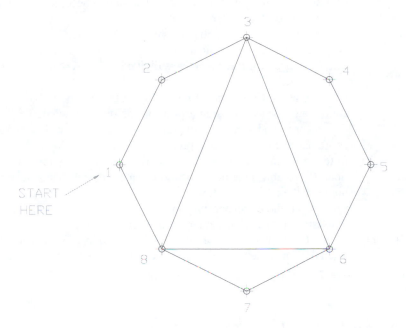

Figure 2-9: Completed line.dwg

Creating Circles and Arcs

There are several methods available to specify the location and form of an arc or circle. This also applies to circles and arcs created with AutoCAD. To draw circles and arcs properly, you must understand the options that are a part of the CIRCLE and ARC commands. These commands and options are discussed in the following sections.

Using the CIRCLE Command

The CIRCLE command has many options. These include center, radius; center, diameter; two points; three points; tangent, tangent, radius; and tangent, tangent, tangent. The following table describes the use of the CIRCLE command and its options.

Methods for invoking the CIRCLE command include:

▶ Toolbar: Draw

▶ Menu: Draw > Circle

▶ Command: CIRCLE

Circle Command Options	Circle Creation Methods
Center, Radius	AutoCAD prompts you by default, to enter a center point. Then you are prompted to enter a radius; you can enter a value for the distance or drag the circle to a desired size. The radius becomes the default for future CIRCLE commands.
Center, Diameter	Select the CIRCLE command center default option. Enter **d** for diameter; then enter a distance or drag the circle to a desired size. The value used for the diameter is divided in half, creating a default radius value.
2 Points	Enter **2p** at the Command prompt. Select the first point by entering coordinates, or by choosing a point in the drawing window. Select the second endpoint by dragging the circle, or by entering coordinates at the Command prompt.
3 Points	Enter **3p** at the Command prompt; then enter the points one at a time. To establish the point locations, enter coordinates or select the points in the display window.
Tangent, Tangent, Radius	Enter **ttr** at the command prompt. Select the first, then the second, line, circle, or arc. Then enter a value for the radius.
Tangent, Tangent, Tangent	Select Circle from the Draw menu, then choose Tan, Tan, Tan. Then draw a circle using tangenial points of three selected objects.

Table 2-1: The Circle command options

Graphic Examples of the Use of Different Options

The following section displays graphic examples of how points are placed when the CIRCLE command options are used to draw a circles. The examples include: (**1**) Center Radius (**2**) Center, Diameter (**3**) 2Point (**4**) 3Point (**5**) Tan, Tan, Radius and (**6**) Tan, Tan, Tan.

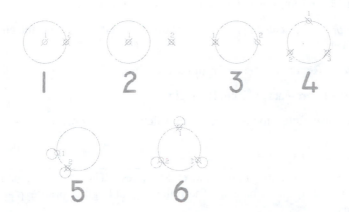

Figure 2-10: Graphic Examples of the Use of Different Options

Using the ARC Command

An arc is a curved line or a partial circle created with the ARC command. Points for the ARC options are specified by selecting random points in the drawing window, or by entering coordinates.

Methods for invoking the ARC command include:

▶ Toolbar: Draw

▶ Menu: Draw > Arc

▶ Command: ARC

There are a number of ways to create arcs. The 3 Points arc option is the default for the ARC command. To draw a 3 Points arc, select a starting point in the drawing window. You are then prompted to enter a second point. Drag the arc to a desired position. The third point represents the endpoint of the arc you are creating. To select the endpoint, press the left mouse button. You may also enter absolute coordinates to located the points of the arc. The ARC default Command prompts are as follows:

```
Center/ <Start point>:
Center/End/<Second Point>:
Endpoint:
```

Arcs can be drawn by using the other ARC command options. A brief explanation of these options follows:

▶ *Start* - Represents the starting point of the arc or the first point of the arc.

▶ *Center* - Represents the center of the arc.

◗ *End* - Represents the endpoint or the last point on the arc.

◗ *Length* - Represents the length of a chord or line that connects the endpoints of an arc.

◗ *Radius* - Represents the radius of the arc.

◗ *Angle* - Represents the included angle.

◗ *Direction* - Refers to the direction in which the arc is drawn.

Arc Options

The ARC command lets you specify options and enter values. The ARC command options are selected from the Draw menu, Arc submenu. The following table gives brief explanations of the ARC command options.

Arc Command Options	Arc Creation Methods
Start, Center, End	When you specify the start and center points, the radius is automatically established and displayed on the display window. The endpoint option then determines the length of the arc
Start, Center, Angle	When you select this option from the Arc menu, you are asked to enter the start point, and then the center point. After these points are determined, enter the value for the angle. An arc drawn counterclockwise will be displayed. Use negative values for the angle if you want it to be drawn clockwise.
Start, Center, Length	Chord lengths to are used to determine the endpoint of the arc. By default, arcs are always drawn counterclockwise. This means a positive value for the length chords creates a minor angle, and a negative value creates a major angle.
Start, End, Angle	Select the starting point and end point of the arc. Then enter the angle value to display the arc.
Start, End, Direction	Select the starting point and the endpoint of the arc, then enter a direction value for the arc direction. The location and size of the arc are then determined by the location of the two points, and the entered direction value.
Start, End, Radius	Arcs are created by establishing points using the Start and End options, then entering a value for the radius. These arcs are also drawn counterclockwise, therefore, positive radius values create small arcs and negative values create large arcs.
Center, Start, End	This option works the same way as the Start, Center, End option except you select the center point first. Use this sequence if you know the location of the arc center point.
Center, Start, Angle	This sequence is a variation of the Start, Center, Angle option. Use this option if it is easier to draw the arc by establishing the center point.
Center, Start, Length	This is a variation of the Start, Center, Length option. Use this option if it is easier to draw your arc from the center point

Continue

instead of from the starting point.

This option draws an arc tangent to the last line or arc drawn. To invoke this option, select Arc from the Draw menu, then choose Continue.

Table 2-2: The Arc command options

Exercise 2-2: Drawing Circles and Arcs

In this exercise, you draw two lamp symbols using the diameter and radius options of the CIRCLE command. You will also create a stool using the CIRCLE command. You will use the ARC command to create a conference table, and two different types of chairs.

Using the CIRCLE Command

1.   Open the file *furni.dwg*. The drawing looks like the following figure:

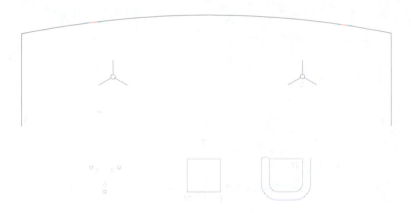

Figure 2-11: The furni.dwg

2.   From the Draw menu, choose Circle, then choose Center, Diameter. In response to the center point, press the F9 key to turn on the Snap mode, if its not already toggled on. Position the cursor at the center of the circle next to the number 1, then press the left mouse button.

3.   For the diameter size, enter **4** then press ENTER.

4.   Press ENTER to repeat the last command. In response to the `center point` prompt, position the cursor at the center of the circle next to the number 2, then press the left mouse button.

5.   The default command option is `<radius>`. Enter **2** for the radius size, then press ENTER.

6.   Use the VIEW command to restore the view named STOOL, then set the current layer to CHAIR.

7.   From the Draw menu, choose Circle, then choose Tan, Tan, Tan. Press the F9 key to turn the Snap mode off, or double click the SNAP button in the status bar.

8.   In response to the first Command prompt, `First point: _tan to`, position the cursor on the edge of the circle closest to the number 3, then press the left mouse button. Repeat the process for numbers 4 and 5. The drawing looks like the following figure:

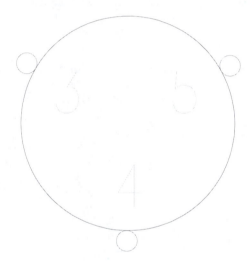

Figure 2-12: A stool created with the Circle, Tan, Tan, Tan option

Making a 3 Points Arc

1.   Press the F9 key to turn the Snap mode on, make the current layer TABLE, then set the current view to All.

2.   From the Draw menu, choose Arc, then choose 3Points.

3.   In response to the Command prompt `<Start point>`, position the cursor at the bottom of the red line next to number 6, then press the left mouse button.

4.   For the `<Second point>`, position the cursor over the number 7, then press the left mouse button.

5.   For the `End point` prompt, position the cursor at the bottom of the red line next to number 8, then press the left mouse button.

Using the ARC Command

1. Make the current layer CHAIR. From the Draw menu, choose Arc, then choose Start, End, Radius. In response to the `Start  point` prompt, position the cursor at the corner next to number 9, then press the left mouse button.

2. For the second point, position the cursor next to number 10, then press the left mouse button.

3. Now you need to enter a value for the radius. Observe that the arc changes as you move the cursor around the drawing window, then enter **3.6** for the radius.

4. From the Draw menu, choose Arc, then choose Start, End, Angle. In response to the `Start  point` prompt, position the cursor at the end of the line next to number 11, then press the left mouse button.

5. For the `second  point` prompt, position the cursor at the corner next to number 12, then press the left mouse button.

6. Now you need to provide an Angle. Enter **90,** then press ENTER.

7. This completes this exercise. The drawing should look like the following figure:

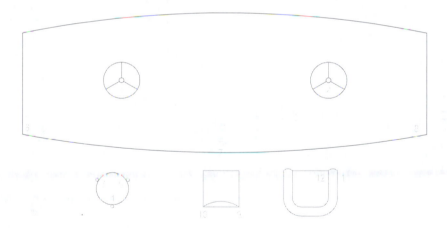

Figure 2-13: The completed furni.dwg.

Conclusion

After completing this chapter, you have learned the following:

▶ You can use the NEW command to specify a filename before you start working.

▶ You can access existing drawing files by using the OPEN command.

- ▶ The SAVE command options let you protect your work by storing the existing status to a named file in a directory.

- ▶ The display commands are used to alter and magnify drawing views.

- ▶ The LINE command and features are used to create objects.

- ▶ Use the CIRCLE and ARC commands, and command options to draw circular and curved objects.

Chapter 3

Creating, Accessing, and Saving AutoCAD Drawings

In this chapter, you learn how new drawing files are created, how to open existing drawings, how to save changes to existing drawings, and how to exit AutoCAD.

About This Chapter

In this chapter, you will do the following:

▶ Use the Drawing Start Up dialog box.

▶ Create a new drawing using Quick and Advanced Setup Wizards.

▶ Create a drawing using a template file.

▶ Save a drawing.

▶ Exit a drawing session.

▶ Open an existing drawing.

▶ Set the automatic timed save feature.

▶ Optimize the creation of backup files.

The Drawing Start Up Dialog Box

When you start an AutoCAD session, the Start Up dialog box is displayed, as shown in the following figure. This dialog box provides you with an easy to use interface when starting an AutoCAD session. Drawings can be created using Quick or Advanced Setup Wizards, Use a Template file, or Start from Scratch. You can open existing drawings using Open a Drawing.

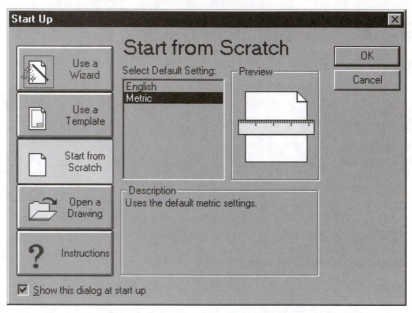

Figure 3-1: The Start Up dialog box

Using Drawing Wizards - Quick and Advanced

The Use a Wizard options lead you through the basic steps of setting up a drawing using either Quick Setup or Advanced Setup. When you choose Use a Wizard, the Quick Setup and Advanced Setup options are displayed, as shown in the following figure:

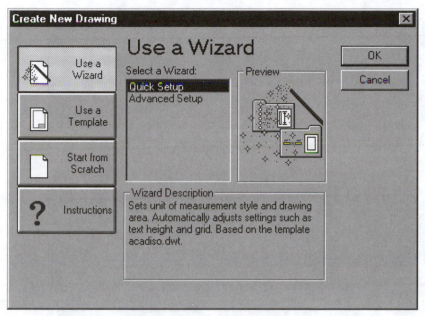

Figure 3-2: The Use a Wizard options

Quick Setup

The Quick Setup wizard leads you through the setup for Units and Area, as shown in the following figure:

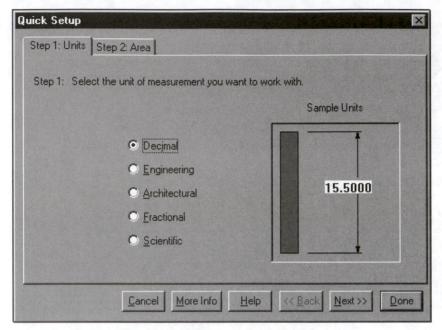

Figure 3-3: The Quick Setup options

The Step 1: Units tab displays five options for the unit of measurement. When you choose an option, AutoCAD displays an example in the Sample Units image tile. The units you choose control how AutoCAD interprets coordinate entries and how it displays coordinates.

The Sample Units image tile is shown in the following figure:

Figure 3-4: Example of units

Examples of each option are shown in the following table:

Unit of Measurement	Example
Decimal	15.5000
Engineering	1'-3.5000"
Architectural	1'-3 ½"
Fractional	15 ½
Scientific	1.5500E+01

Table 3-1: Examples of units of measurement

To obtain text-based help on the menu equivalent for the active tab, choose the More Info button. An Information box is displayed, as shown in the following figure:

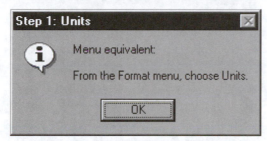

Figure 3-5: The More Info dialog box

You can move to the Step 2: Area tab by choosing the tab itself, or by choosing the Next button. The Step 2: Area tab is displayed, as shown in the following figure:

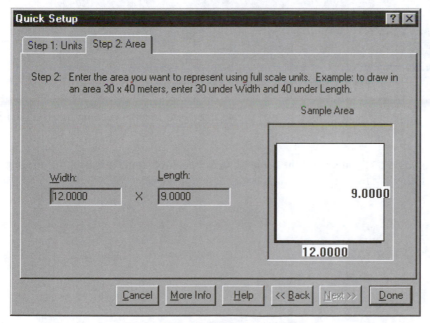

Figure 3-6: The Step 2: Area tab

Based on the limits of your drawing, you can enter suitable values for width and length in their respective fields. The Sample Area image tile displays an example of your drawing limits using the units setting from Step 1 and the values you entered for width and length. To complete the setup, choose the Done button. The AutoCAD drawing window displays the limits and grid established by the values selected in the Quick Setup Wizard.

Advanced Setup

The Advanced Setup wizard leads you through the setup for Units, Angle, Angle Measure, Angle Direction, Area, Title Block, and Layout, as shown in the following figure:

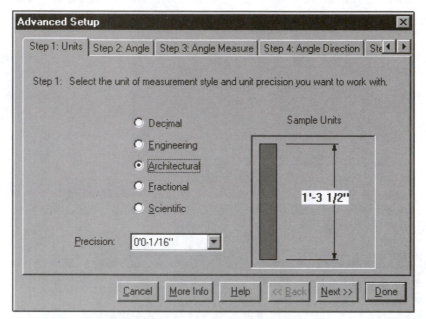

Figure 3-7: The Advanced Setup Wizard

In the Step 1: Units tab, you can set the coordinate interpretation and readout as discussed in the previous section, "Quick Setup".

In the Step 2: Angle tab, you can set the angle of measurement and the precision for the angles, as shown in the following figure:

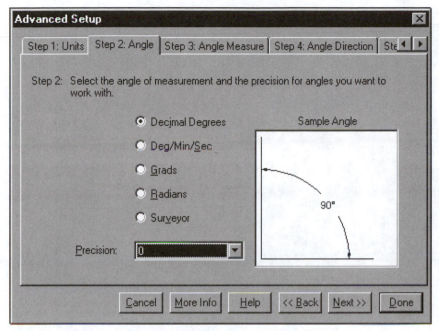

Figure 3-8: The Step 2: Angle tab

Examples of the five angle options are shown in the following table:

Angle of Measurement	Example
Decimal Degrees	90°
Deg/Min/Sec	90d
Grads	100g
Radians	2r
Surveyor	N 0d W

Table 3-2: Examples of angle of measurement

In the Step 3: Angle Measure tab, you can select the direction from which AutoCAD measures angles. The five options are shown in the following figure:

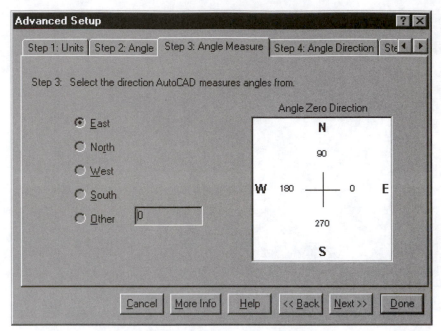

Figure 3-9: The Angle Measure tab

In the Step 4: Angle Direction tab, you can select the direction in which AutoCAD measures angles. The two options, Clockwise and Counter-Clockwise, are shown in the following figure:

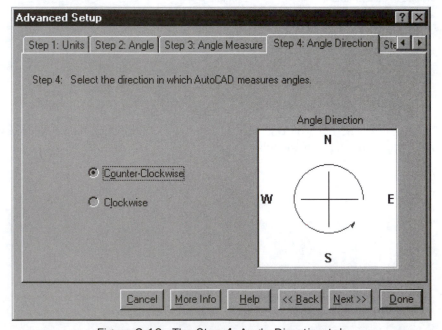

Figure 3-10: The Step 4: Angle Direction tab

In the Step 5: Area tab, you can set the drawing limits, as discussed in the "Quick Setup" section. The Step 6: Title Block tab gives you the option of inserting a pre-defined title block in your drawing. A title block consists of a pre-defined border which conforms to national standards, and an area for text-based information such as company name, scale, and drawing number. Title blocks for ANSI, DIN, ISO, and JIS standard sheet sizes are provided. Customized title blocks can be added by choosing the Add button.

To select a title block, you select from the Title Block Description and Title Block File Name drop-down lists, as shown in the following figure:

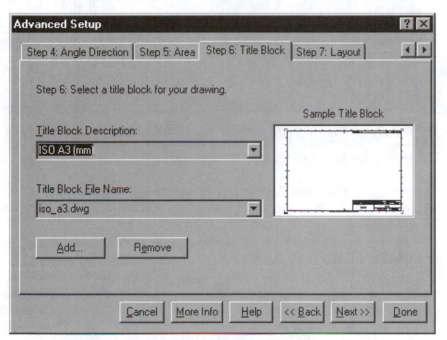

Figure 3-11: Selecting a title block

The Step 7: Layout tab gives you the option of using the advanced paper space layout capabilities. If you select Yes, you have three options to select from, as shown in the following figure:

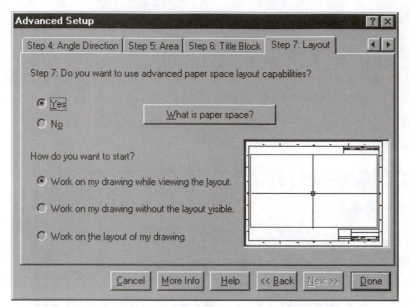

Figure 3-12: The Step 7: Layout tab

If you select No, the options under How Do You Want to Start? are not available.

When you have completed the seven steps in Advanced Setup, choose Done to complete the process.

Using Template Files

Conformance to standards is critical in the design process, ensuring consistent application of national and local practices, and the effective, productive use of AutoCAD. *Template drawings* are drawings with pre-defined settings based upon the required standards for your current project. This means that all users will produce drawings that are consistent and adhere to national and local standards. These files are similar in use to the templates in Microsoft Office.

The Use a Template option lets you create a new drawing based on a template drawing. All template drawings have a *.dwt* file extension. Selecting Use a Template displays an alphabetical list of the available template files, as shown in the following figure:

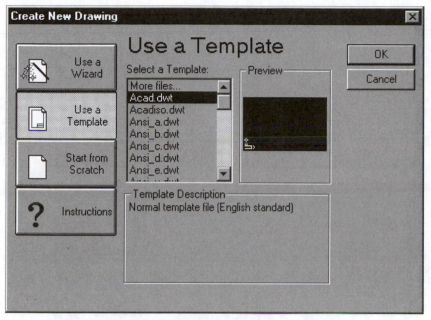

Figure 3-13: The Use a Template option

The template files are stored in a template sub-directory, normally within the main AutoCAD directory. This location is set by choosing Template Drawing File Location on the Files tab of the Preferences dialog box, as shown in the following figure:

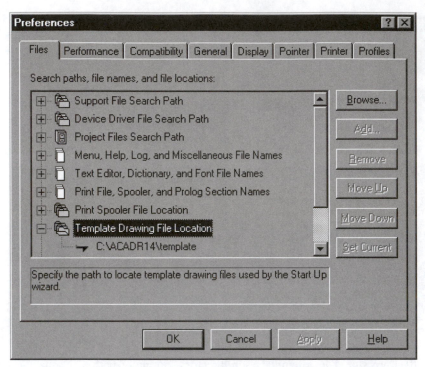

Figure 3-14: Setting the Template Drawing File location

Template files are the equivalent of prototype drawings in previous releases of AutoCAD. By contrast, prototype drawings have a *.dwg* extension, while template files have a *.dwt* extension that helps protect the file from inadvertent changes.

The default template file is *acad.dwt* or *acadiso.dwt*, depending upon initial installation of the AutoCAD software. Any drawing file can be saved as a template file. However, given the importance of setting standards for any drawing session, it is recommended that template files be created prior to the start of a project. Existing prototype drawings can be saved as template files, thereby ensuring that previous work of creating standard drawing files will be maintained.

AutoCAD supplies a number of template files, some of which are listed in Figure 3-13. If the drawing file you want to use is not listed, you can select More Files from the Select a Template drop-down list, then choose OK. This displays the Select Template dialog box, as shown in the following figure:

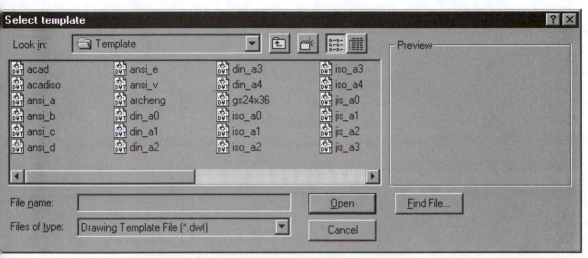

Figure 3-15: The Select Template dialog box

You can now access any *.dwt* or *.dwg* file and use that file as a template for your current session. To list either *.dwt* or *.dwg* files you must select from the Files of Type drop-down list, as shown in the following figure:

Figure 3-16: The Files of Type drop-down list

Start from Scratch

The Start from Scratch option uses basic default settings based on *acad.dwt* and *acadiso.dwt*. These are pre-defined settings and are listed as English and Metric in the Create New Drawing dialog box, as shown in the following figure:

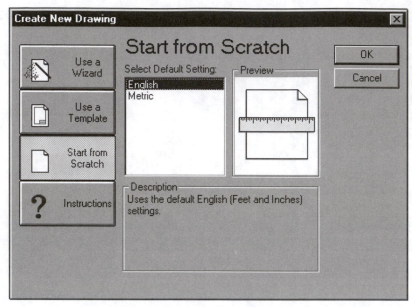

Figure 3-17: The Start from Scratch option

Instructions

Selecting the Instructions option from the Create New Drawing dialog box displays text-based help on the uses of Use a Wizard, Use a Template, and Start from Scratch, as shown in the following figure:

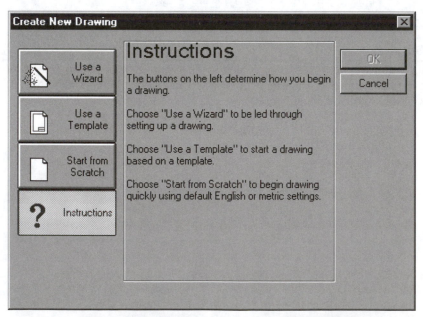

Figure 3-18: The Instructions text-based help

As you can see from the previous figure, the Open a Drawing option is not available for selection. This option is only available at the beginning of an AutoCAD session. Starting a new drawing using any of the options discussed in this section displays the Create New Drawing dialog box without the Open a Drawing option. Opening an existing drawing is discussed later in this chapter.

Saving a Drawing

When you create a new drawing in AutoCAD, the filename default is *drawing.dwg*. Use the SAVE command to save the drawing using another name. AutoCAD automatically adds the *.dwg* extension to the drawing filename you enter. Other options are the QSAVE, SAVEAS, and EXIT commands. To start a new drawing without saving the current drawing, use the QUIT command.

Using the SAVE Command

When you want to save an unnamed drawing which is currently using the default *drawing.dwg* filename, you can enter **save** at the Command prompt.

> Note: You can only use the SAVE command at the Command prompt. The QSAVE command, which is discussed in the next section, is generally used for the first save, with the SAVE command being used to create a new incremental file.

This will display the Save Drawing As dialog box, as shown in the following figure:

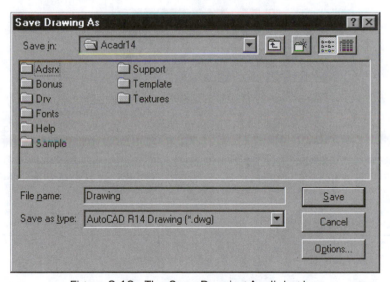

Figure 3-19: The Save Drawing As dialog box

You can then enter a filename in the File Name field, and then choose the Save button. Your current drawing will be displayed and the filename in the title bar at the top of the screen will change from Drawing to the filename you entered.

If you enter **save** after you have named the drawing, the Save Drawing As dialog box will be displayed. If you do not change the filename and choose Save, an Alert box will be displayed, as shown in the following figure:

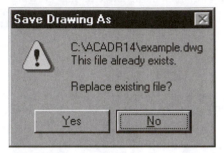

Figure 3-20: Save Drawing As Alert box

Choosing Yes saves the most current version of your drawing which is then displayed. Choosing No returns you to the Save Drawing As dialog box. You can then save the drawing with a different name. However, there are more effective methods of doing this, which are covered in the QSAVE and SAVEAS sections.

The Save As Type drop-down list displays four options for saving the current drawing. These options are shown in the following figure:

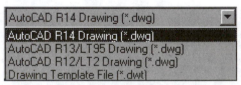

Figure 3-21: The Save As Type options

The default option is to save the file as an AutoCAD R14 drawing. You can also save the drawing as an AutoCAD R13 or LT95 drawing for backward compatibility. Similarly, you can save the drawing as an AutoCAD R12 or LT2 drawing. If the current drawing was created to establish standards for a specific project you can save the file using the Drawing Template File option.

Using the QSAVE Command

The SAVE command is only available from the Command prompt. When you choose Save from the File menu or the standard toolbar, you get the QSAVE command.

Methods for invoking the QSAVE command include:

▶ **Toolbar:** Standard

▶ **Menu:** File > Save

▶ **Command:** QSAVE

If your file is unnamed, the Save Drawing As dialog box is displayed. If your drawing is named then AutoCAD saves the file using the current filename without displaying a dialog box. This is a more efficient method of saving a file over the SAVE command discussed in the previous section.

Using the SAVEAS Command

You may want to save your current drawing using a different filename, or as a different filetype using the filetypes listed in the Save As Type drop-down list. To do this, use the SAVEAS command.

Methods for invoking the SAVEAS command include:

- ▶ Menu: File > SaveAs

- ▶ Command: SAVEAS

If your current drawing is named, the filename you enter becomes the new current filename. Also, if your current drawing is unnamed, the filename you enter becomes the new current filename.

Using the EXIT Command

When you have finished creating and editing your current drawing, and you want to exit AutoCAD, use the EXIT command. Methods for invoking the EXIT command include:

▶ **Menu:** File > Exit

▶ **Command:** EXIT

If you have any unsaved changes to your drawing, an Alert box will be displayed, as shown in the following figure:

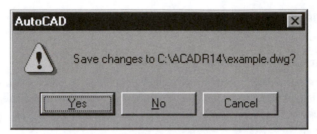

Figure 3-22: The Alert dialog box

Choosing the Yes button saves the changes to the current filename and exit AutoCAD. If the file is unnamed, the Save Drawing As dialog box is displayed. Choosing No exits AutoCAD without saving the changes to your drawing since the most recent save. If you decide not to exit and want to return to AutoCAD, choose Cancel.

Using the QUIT Command

When you have saved the most recent changes to a file and want to quit AutoCAD, use the QUIT command. Methods for invoking the QUIT command include:

▶ Menu: File > Exit

▶ Command: QUIT

If there are any changes which have not been saved, an Alert box is displayed as shown in the previous figure. You can then decide if you want to save the changes or QUIT without modifying the current drawing.

> Note: EXIT and QUIT are functionally the same. Redundancy is to maintain compatibility with previous versions.

Opening an Existing Drawing

You can open an existing drawing when you start a new AutoCAD session, or while you are working on another AutoCAD drawing. You must save or close without saving the drawing you are currently working on in order to open a previously saved drawing.

Select File Dialog Box

When you start a new AutoCAD session, the Open a Drawing option is available on the Start Up dialog box, as shown in the following figure:

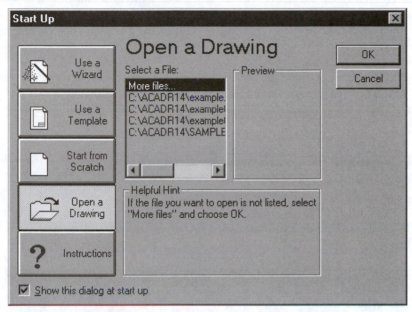

Figure 3-23: The Open a Drawing option

When you select Open a Drawing, the Select a File list box is displayed, as shown in the following figure. This list box lists an entry named More Files, and the names of the last four opened drawings, matching the drawing history displayed in the File pull-down menu. If the drawing you want to open is not listed you can select More Files, then choose OK. The Select File dialog box is displayed, as shown in the following figure:

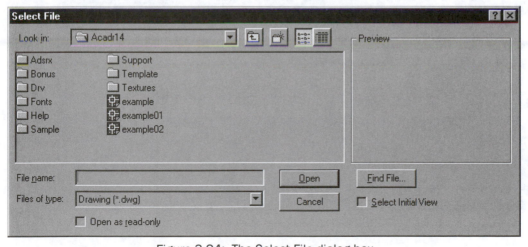

Figure 3-24: The Select File dialog box

The current folder is displayed in the Look In drop-down list. The sub-folders and drawing filenames in the current folder are listed. When you select a filename, the Preview image tile displays an image of the objects in the drawing. If this is the drawing you want to open, choose the Open button and AutoCAD displays the drawing. You can also double-click on the file name to open the drawing.

If you want to list template files with a *.dwt* file extension, you can select Drawing Template File (*.dwt*) from the Files of Type drop-down list, as shown in the following figure:

Figure 3-25: The Files of Type drop-down list

You use the Find File feature to view images of drawings, open drawings, and search for files. When you choose the Find File button, the Browse/Search dialog box is displayed, as shown in the following figure:

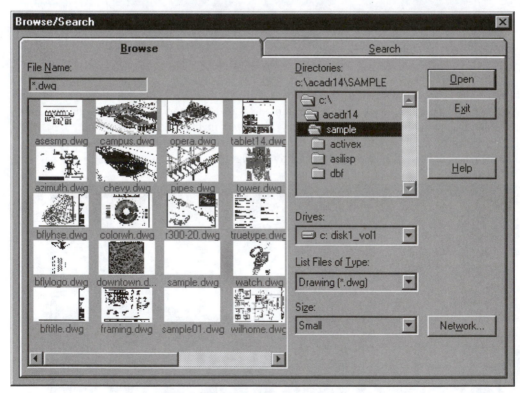

Figure 3-26: The Browse/Search dialog box

You use the Browse tab to view images in a specified drive and file type. The directory structure is displayed and you can then select a folder. The size of the image can be set to small, medium, or large by selecting from the Size drop-down list. To open a file, select the image, then choose the Open button. You can also double-click on the image to open the drawing.

The Search tab, shown in the following figure, lets you search for files by specifying a search pattern based on the selected file type, referencing their date of creation, and a search location.

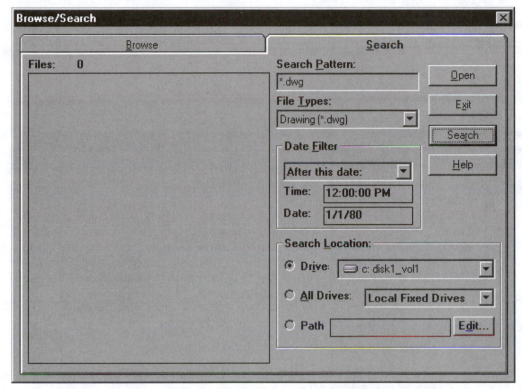

Figure: 3-27: The Search option

When you have entered the search pattern, choose the Search button. This button then changes to Stop Search, letting you to stop the search at any time. The files which match the pattern are listed with a full path and image of the drawing. To open a drawing, select the image and then choose the Open button. You can also double-click on the image to open the drawing.

The methods of opening an existing drawing discussed in the previous section can only be used if you are opening a file at the beginning of an AutoCAD session. To open an existing file during an AutoCAD session, you use the OPEN command.

Methods for invoking the OPEN command include:

▶ Toolbar: Standard

▶ Menu: File > Open

◗ Command: OPEN

The OPEN command displays the Select File dialog box, as shown in Figure 3-24.

During a design session, the NEW command will display the Create a New Drawing dialog box. All options from Start Up are available, except Open a Drawing.

Methods for invoking the NEW command include:

◗ Toolbar: Standard

◗ Menu: File > New

◗ Command: NEW

Exercise 3-1: Creating and Saving Drawing Files

 Starting an AutoCAD session displays the Start Up dialog box. The options on this dialog box let you use either the Quick or Advanced Wizards, Use a Template file, Start from Scratch, or Open a Drawing.

In this exercise, you start a new drawing using the Quick and Advanced Wizard options. These wizards will guide you through the initial setup of a drawing. You then create a template drawing file and use the file to start a new drawing. This will illustrate how you can use template files to set standards for all drawings. Having created a new drawing, you then save the current drawing and exit AutoCAD.

Using Start from Scratch

1. Start a new AutoCAD session. The Start Up dialog box is displayed.

    Note: By default, the dialog box will display the option selected to start the previous AutoCAD session.

2. Choose the Instructions button. Text-based help is displayed.

3. Choose Start from Scratch, then select Metric, as shown in the following figure:

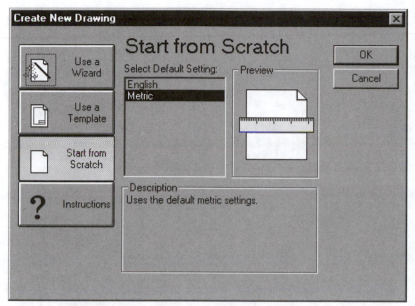

Figure 3-28: Starting a drawing from scratch

4. Then choose OK. The default filename is *drawing.dwg* and it is based on *acadiso.dwt*. Move the cursor around the drawing window and note the coordinate display in the status bar. The width and length of the drawing are set to 420 x 297 units.

In the next part of the exercise you are guided through the creation of a new drawing using the Quick Wizard option.

Using the Quick Setup Wizard

1. From the File menu, choose New. AutoCAD displays the Create New Drawing dialog box. Note that the Open a Drawing option is not available. This option is displayed only on the Start Up dialog box.

2. Choose the Use a Wizard option. In the Select a Wizard list box, select Quick Setup. Then choose OK. The Quick Setup dialog box is displayed, as shown in the following figure:

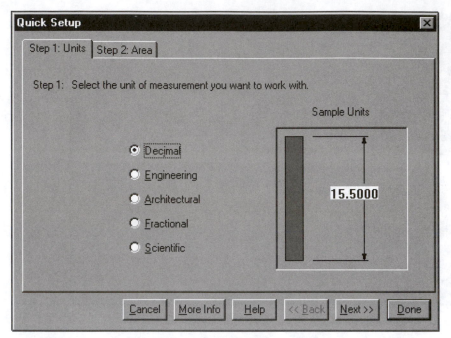

Figure 3-29: The Quick Setup dialog box

3. In the Step 1: Units tab, the default is Decimal. Choose the Step 2: Area tab, or choose the Next button to advance to the next step.

4. Enter **297** for Width, and **210** for Length, as shown in the following figure.

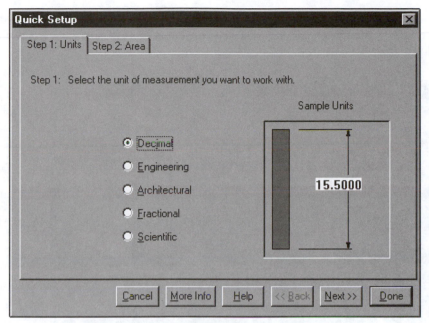

Figure 3-30: Setting the drawing area

5.   Then choose the More Info Button. An Alert box is displayed, as shown in the following figure.

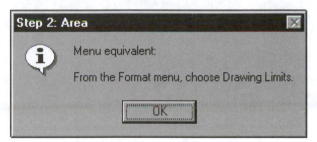

Figure 3-31: The Area Alert box

6.   Choose OK, then choose Done to complete the setup. AutoCAD displays the drawing limits and grid established by the specified settings.

The Quick Wizard guides you through two steps and requests a minimum amount of information. The Advanced Wizard is more suitable for the experienced user and gives you a greater degree of control in setting up your drawing environment.

Using the Advanced Wizard

1.   From the File menu, choose New. Do not save changes to *drawing.dwg.*

2. Choose the Use a Wizard option. In the Select a Wizard list box, select Advanced Setup. Then choose OK. The Advanced Setup dialog box is displayed, as shown in the following figure:

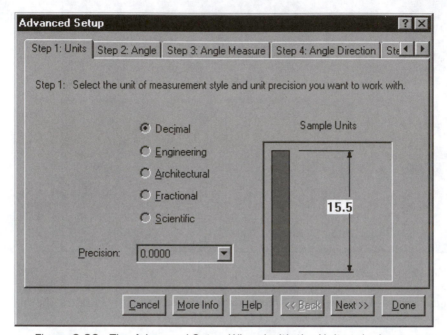

Figure 3-32: The Advanced Setup Wizard with the Units tab chosen

3. In the Step 1: Units tab, set Precision to two decimal places in the Precision drop-down list. Move to the next tab by choosing Next.

4. In the Step 2: Angle tab, select Decimal Degrees and set Precision to 1 decimal place, as shown in the following figure:

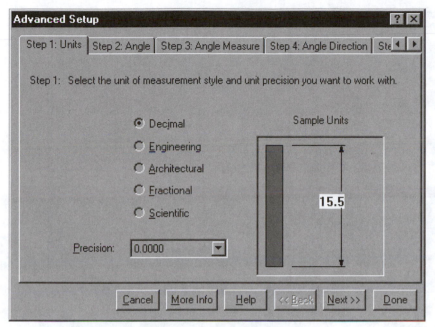

Figure 3-33: The Step 2: Angle tab

5. The Step 3: Angle Measure tab and the Step 4: Angle Direction tab will use the default values.

Move to the Step 5: Area tab, as shown in the following figure.

In the Step 5: Area tab, set the area to **594** for Width and **420** for Length. Move to the next tab.

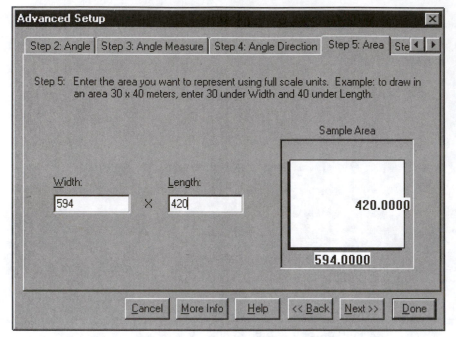

Figure 3-34: The Step 5: Area tab

6. In the Step 6: Title Block tab, in the Title Block Description drop-down list, select the ISO A2 title block, as shown in the following figure. Move to the next tab.

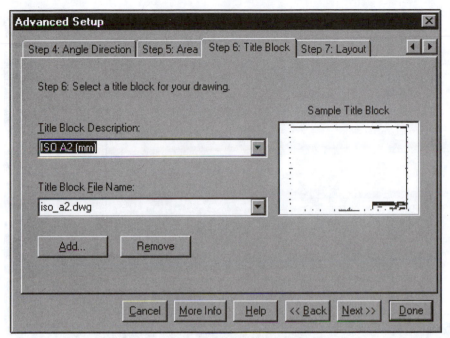

Figure 3-35: The Step 6: Title Block tab

7. In the Step 7: Layout tab, keep the default settings which use an advanced paper space layout and let you work on the drawing while viewing the layout. Then choose Done.

    AutoCAD opens a new drawing in paper space with a single floating viewport, and the title block inserted, as shown in the following figure:

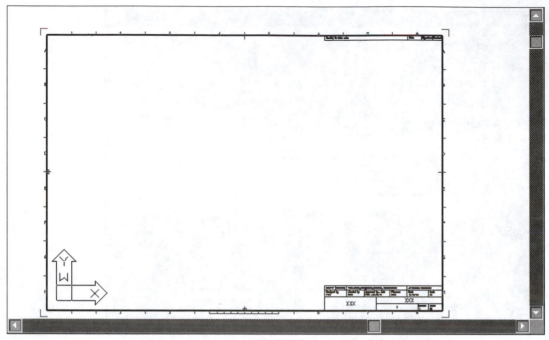

Figure 3-36: Completed drawing with title block

Using a Template File

1.	From the File menu, choose New. Do not save changes.

2.	Choose Use a Template. Then select *iso_a2.dwt* from the Select a Template list of files, as shown in the following figure:

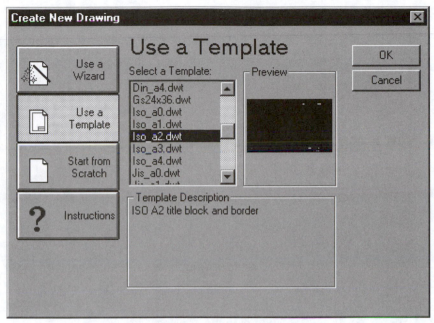

Figure 3-37: Selecting iso_a2.dwt

A preview of the file is displayed in the Preview image tile, and a brief description is listed in the Template Description area. Then choose OK.

The ISO A2 border is displayed in the drawing window, as shown in the following figure:

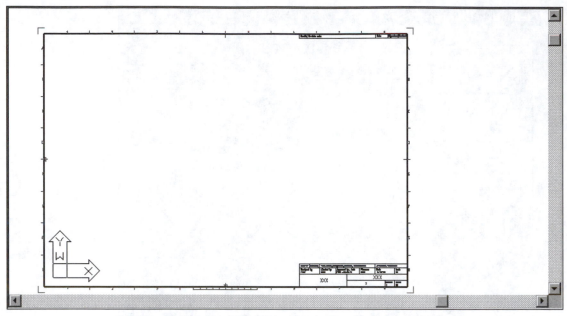

Figure 3-38: The ISO_A2 border in the drawing window

You are now ready to start working on your drawing. The name of the current file is *drawing.dwg*. To save the file using another name, you use the SAVE command.

Using the SAVE Command

1. At the Command prompt, enter **save** and press ENTER. The Save Drawing As dialog box is displayed. In the File Name field, enter **exer01**, as shown in the following figure:

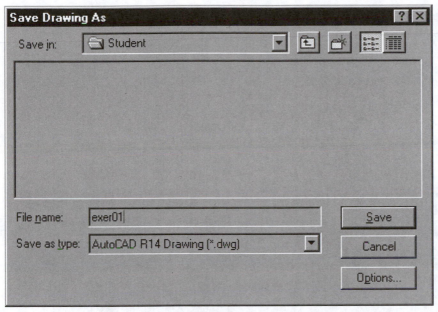

Figure 3-39: Saving the file as exer01.dwg

Then choose the Save button. The drawing filename is changed to *exer01.dwg*.

As your work in this drawing progresses you will be creating valuable data. To ensure that this data is saved on a regular basis you can set the automatic timed save interval to a suitable value.

Setting Automatic Timed Save

The system variable SAVETIME sets the time interval in minutes for the automatic timed save.

1. From the Tools menu, choose Preferences. Then choose the General tab. Make sure that the Automatic Save checkbox is checked, and enter a value of **30** in the Minutes Between Saves field, as shown in the following figure:

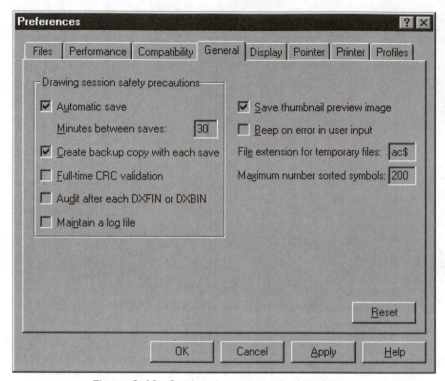

Figure 3-40: Setting the automatic timed save

Then choose OK.

If you work in this drawing for 30 minutes your drawing will be automatically saved as *acad.sv$*. Performing a manual save with SAVE, SAVEAS, or QSAVE will reset the timer.

Exiting AutoCAD

When you have completed working on your drawing, you can start a new drawing using any of the techniques you used in this exercise. If you want to exit AutoCAD you can use either the EXIT or QUIT command. In both commands, if you have work that has not been saved, you will be prompted to save the changes before exiting.

1. From the File menu, choose Exit. If you have any unsaved changes the Alert box will be displayed, as shown in the following figure:

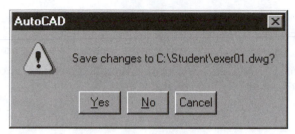

Figure 3-41: The Alert box

Choose Yes. You will exit from your current AutoCAD session. If there were no changes to your drawing, AutoCAD will exit without any further prompts.

Exercise 3-2: Opening an Existing Drawing

The previous exercise demonstrated the options for creating a new drawing. In this exercise, you open existing files that you want to modify or view. You use the OPEN command and the options in the Select File dialog box. In this Explorer-type dialog box, you use the search capability in Find File.

Using the Select File Dialog Box

1. Start a new AutoCAD session.

2. Select Start from Scratch, then select Metric. Then choose OK.

3. From the File menu, choose Open. The Select File dialog box is displayed, as shown in the following figure:

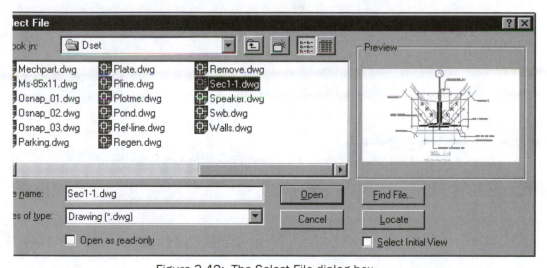

Figure 3-42: The Select File dialog box

Note: Select File is also available from Open a Drawing, at the beginning of any new AutoCAD session. Select More Files from the Select a File list box, then choose OK to display the Select File dialog box.

4. Select *sec1-1.dwg*. Then choose Open. A section view of a column footing is displayed, as shown in the following figure:

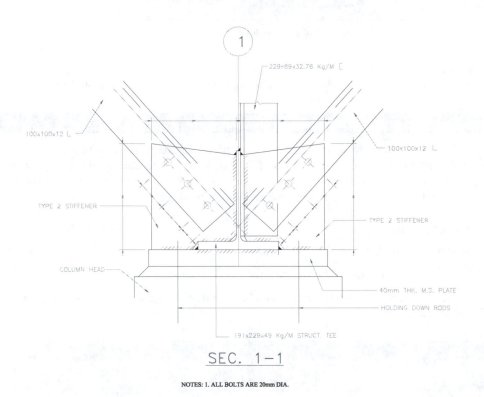

SEC. 1–1

NOTES: 1. ALL BOLTS ARE 20mm DIA.

Figure 3-43: Sec 1-1.dwg

Using the Browse/Search Dialog Box

1. From the File menu, choose Open. The Select File dialog box is displayed.

2. The file you want to open is named *azimuth.dwg*. It is not listed in the current directory. Choose the Find File button. The Browse/Search dialog box is displayed. Choose the Search tab.

3. Enter **az*.dwg** in the Search Pattern, choose All Drives in the Search Location area, then choose the Search button. A list of files which meet your criteria is displayed. One file is found and displayed, as shown in the following figure:

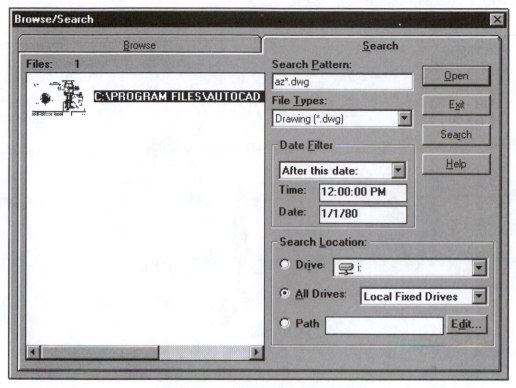

Figure 3-44: The Search tab

4.   Open the drawing file by double-clicking on the preview image. The drawing should look like the following figure:

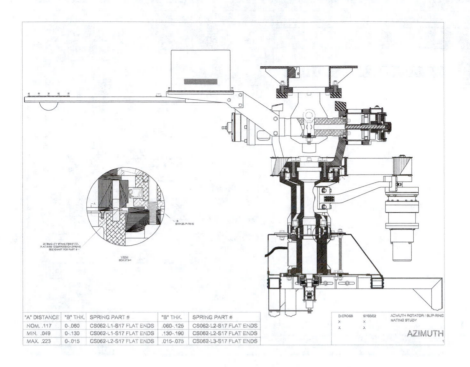

"A" DISTANCE	"B" THK.	SPRING PART #	"B" THK.	SPRING PART #				
NOM. .117	0-.060	CS062-L1-S17 FLAT ENDS	.060-.125	CS062-L2-S17 FLAT ENDS	D.CROSS	97/03/02	AZIMUTH ROTATOR / SLIP-RING MATING STUDY	
MIN. .049	0-.130	CS062-L1-S17 FLAT ENDS	.130-.190	CS062-L2-S17 FLAT ENDS	X	X		
MAX. .223	0-.015	CS062-L2-S17 FLAT ENDS	.015-.075	CS062-L3-S17 FLAT ENDS	X	X	AZIMUTH	

Figure 3-45: Azimuth.dwg

5.   From the File menu, choose Exit.

File Management

The data that you create during a drawing session should be saved on a regular basis in order to minimize loss of data. Creating backup versions is also important in the process of effectively managing data. AutoCAD lets you set an automatic timed save and creates backup copies of your drawings. The backup file is saved with a *.bak* filename extension.

Automatic Timed Save

When you are working on a drawing you may forget to save the drawing on a regular basis. To compensate for this, AutoCAD provides a system variable named SAVETIME which sets an interval in minutes at which point your drawing will be automatically saved. The value can be set at the Command prompt, or on the General tab of the Preferences dialog box.

Methods for invoking the SAVETIME command include:

- ▶ **Menu:** Tools > Preferences > General

- ▶ **Command:** SAVETIME

The General tab of the Preferences dialog box is displayed, as shown in the following figure:

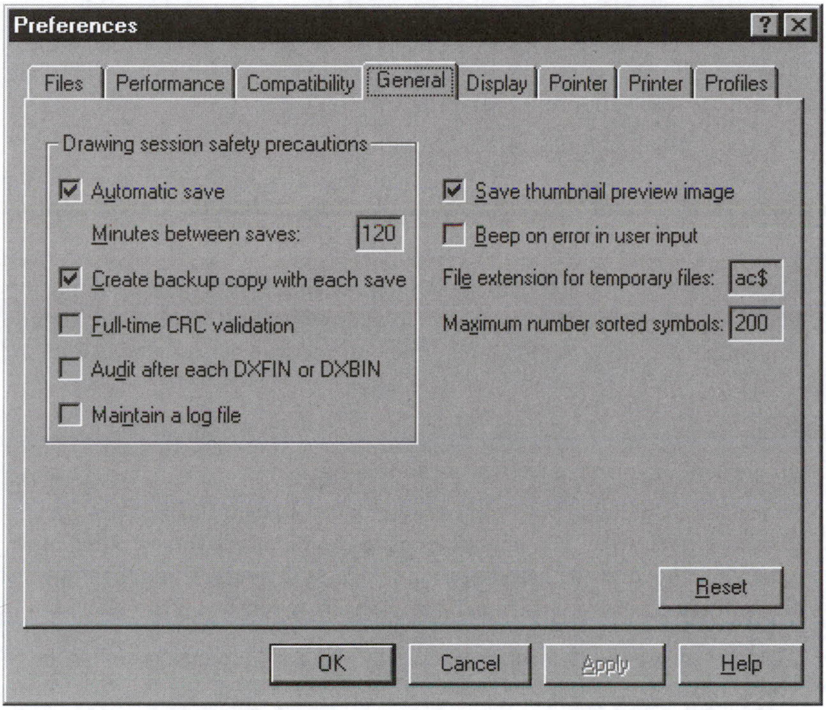

Figure 3-46: Setting the Automatic save value

The default value for Automatic save is 120 minutes. This should be set to a smaller value to protect the data you are creating in your drawing. A value of 30 is a suitable time between automatic saves. The file, by default, is saved as *acad.sv$*.

*.bak*Files

When you save your current drawing and exit AutoCAD, you have created a *.dwg* file. When you subsequently use SAVE or SAVEAS during an AutoCAD session, AutoCAD creates a *.bak* file from the previously saved *.dwg* file.

You use this file as part of a data recovery procedure, should your system fail and the modifications to the current drawing not be saved. The backup file uses the current drawing filename and uses a *.bak* extension. The files are saved to the same directory as the current drawing file. To recover the data you should copy the *.bak* file to a *.dwg* file. This can be done by highlighting the *.bak* file in the Explorer or My Computer in Windows 95 or NT 4.0. Then, from the File menu, choose Rename.

Note: The following system variables are associated with saving a drawing:

- ISAVEBAK - Improves the speed of incremental saves, especially for large drawings by disabling the creation of a backup file. isaveback controls the creation of a backup file (BAK).

▶ ISAVEPERCENT - Determines the amount of wasted space tolerated in a drawing file. The value of ISAVEPERCENT is an integer between 0 and 100. The default value of 50 means that the estimate of wasted space within the file does not exceed 50% of the total file size.

▶ SAVEFILE - Stores current auto-save file name

▶ SAVENAME - Stores the file name and directory path of the current drawing once you save it

▶ SAVETIME - Sets the automatic save interval, in minutes. The savetime timer starts as soon as you make a change to a drawing. It is reset and restarted by a manual SAVE, SAVEAS, or QSAVE. The current drawing is saved to auto.sv$.

Conclusion

After completing this chapter, you have learned the following:

▶ In the Start Up dialog box, you can setup new drawings using the Quick and Advanced Wizards. The Use a Template option offers you an efficient way of starting a new drawing based upon an existing drawing, while the Start from Scratch option lets you choose either English or Metric units.

▶ To open an existing drawing, you can use the Start Up dialog box or the OPEN command.

▶ After creating or modifying a drawing, you can save and name the drawing file using the SAVE, or SAVEAS command.

▶ After creating or modifying a drawing, you can save the drawing file using the QSAVE command.

▶ You can setup the automatic timed save variable, SAVETIME, to save your drawings on a regular basis and protect the data you create.

▶ Upon completion of your drawing you can use either the EXIT or QUIT command to end your current AutoCAD session.

Chapter 4

Drawing Setup and Standards

In this chapter, you learn how new drawing files are created using a template that contains standard settings, and how to customize the settings to suit the needs of the project you are working on. You will also calculate the size of the drawing area and use drawing aids to improve your productivity and drawing precision.

About This Chapter

In this chapter, you will do the following:

▶ Establish drawing units using the Units Control dialog box.

▶ Set drawing limits based upon the required drawing scale.

▶ Use the Drawing Aids dialog box to set snap, grid, and drawing modes.

Establishing Drawing Units

In AutoCAD you determine what the unit type and displayed precision will be based on the project requirements. When you have determined what the units and precision should be you can set the values using the Units Control dialog box, or you can let AutoCAD guide you through the setup using the drawing wizards discussed in Chapter 3 "Creating and Accessing Drawing Files". This chapter will focus on setting up the drawing using the Units Control dialog box to set the drawing units and angles. You will also use the Drawing Aids dialog box to set grid and snap spacing and drawing modes, such as ortho and blipmodes.

Units Control Dialog Box

You can setup the format of the coordinate and angle display and set the precision using the Units Control dialog box, as shown in the following figure:

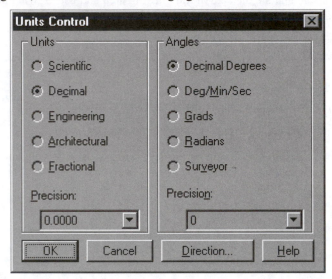

Figure 4-1: The Units Control dialog box

Methods for opening the Units Control dialog box include:

▶ **Menu:** Format > Units

▶ **Command:** DDUNITS

The Units area displays five unit format options and a drop-down list for setting unit Precision. Setting Decimal, Engineering, and Scientific to 4 decimal places of accuracy and setting Architectural and Fractional precision to 1/16 would display the following coordinates:

Unit of Measurement	Example
Scientific	1.5500E+01
Decimal	15.5000
Engineering	1'-3.5000"
Architectural	1'-3 1/16"

Fractional   15 1/2

Table 4-1: Examples of unit format and precision

The Angles area also displays five options and a drop-down list for setting precision. Examples of each option with precision set to 0 decimal places are shown in the following table:

Angle of Measurement	Example
Decimal Degrees	90°
Deg/Min/Sec	90d
Grads	100g
Radians	2r
Surveyor	N 0d W

Table 4-2: Examples of the Angles options

Choosing the Direction button displays the Direction Control dialog box, as shown in the following figure:

Figure 4-2: The Direction Control dialog box

From this dialog box, you can select the direction from which AutoCAD measures angles, either in a *Counter-Clockwise* or a *Clockwise* direction. You can also use the Other option to set a value by selecting two points in the drawing window, or by entering a value at the Command prompt. If you set the East, North, West, and South options to Counter-Clockwise, AutoCAD will measure angles as shown in the following figure:

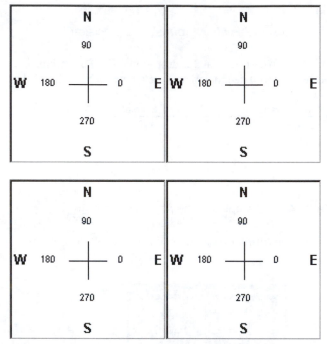

Figure 4-3: Angle 0 direction for East, North, West, and South

When you have selected the appropriate angle and direction for your project, choose OK to exit the dialog boxes.

Note: The following system variables are associated with drawing units:

▶ LUNITS - IS used to set the units using the following values:

  ▶ *1 - Scientific*
  ▶ *2 - Decimal*
  ▶ *3 - Engineering*
  ▶ *4 - Architectural*
  ▶ *5 - Fractional*

▶ LUPREC - Sets the number of decimal places for linear units in commands, (except dimensioning commands) variables, and output.

Setting Drawing Limits

The limits of your drawing are determined by the size of the object you want to draw, the area around the drawing for annotations, and the drawing scale factor. The limits are set in AutoCAD by entering XY coordinates for the *lower-left* and *upper-right* corners of your drawing.

Typically, you will draw the object at a scale of 1:1. The plotting capabilities in AutoCAD let you set an appropriate scale for the final plotted output. This is done using a feature named Paper space which will be discussed in Chapter 16 of this guide. Using this feature, you simply set the limits based on the size of the object you are drawing. However, you may work on a project which requires the drawing to be created using a pre-established scale factor. When that situation occurs, you must follow the process covered in the next section.

Determining the Drawing Scale Factor

Setting a scale factor other than 1:1 requires a calculation based on the actual size of the object and the drawing scale. If you are drawing a floor plan of a house at a scale of 1:50 and you intend to plot the drawing on a sheet size of 210 x 297 then you calculate the *upper-right* limits by multiplying the length and width of the sheet by the second half of the ratio, as follows:

210 x 50 = 10500 mm
297 x 50 = 14850 mm

The limits of drawing are 0,0 and 10500,14850. These limits are using millimeters as the base unit.

Setting Limits Using the LIMITS Command

When you have established what you want the drawing limits to be, you can set the limits using the LIMITS command.

Methods for invoking the LIMITS command include:

- **Menu:** Format > Drawing Limits

- **Command:** LIMITS

The LIMITS command will prompt you for the lower-left coordinates and the upper-right coordinates.

```
Command: '_limits
Reset Model space limits:
ON/OFF/<Lower left corner> <0.0000,0.0000>:
Upper right corner <420.0000,297.0000>:
```

You can also enter the coordinate values by selecting a point in the drawing window. However, entering the values at the Command prompt is the preferred method of setting the drawing limits. It is recommended that you use the All option of the ZOOM command after a change in the limits of your drawing. This displays the area defined by the new limits that you have entered.

The LIMIT command ON option enables limit checking. This prevents you from entering points outside the drawing limits. However, objects such as arcs and circles can extend outside the limits if their center point is placed within the drawing limits. The LIMIT command OFF option will disable limit checking.

> Note: The following system variables are associated with setting the limits of a drawing:
>
> ▶ LIMCHECK - Controls the creation of objects outside the drawing limits. It is the equivalent of setting limits checking to on or off in the LIMITS command.
>
> ▶ LIMMAX - Stores the upper right drawing limits for the current space, expressed as a World coordinate.
>
> ▶ LIMMIN- Stores the lower-left drawing limits for the current space, expressed as a World coordinate.

Setting Snap, Grid, and Drawing Modes

Using the Drawing Aids Dialog Box

AutoCAD provides you with a number of features that will increase your efficiency and make you more productive. Many of these features are contained in the Drawing Aids dialog box. They include Snap, Grid and Ortho, as shown in the following figure:

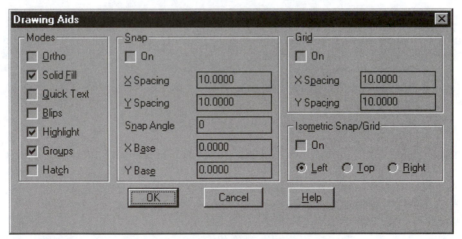

Figure 4-4: The Drawing Aids dialog box

Methods for opening the Drawing Aids dialog box include:

▶ **Menu:** Tools > Drawing Aids

▶ **Command:** DDRMODES

When you are placing a line, arc or circle in your drawing, the easiest method is to enter points by

selecting a point in the drawing window. The problem with this method is that it is inaccurate.
To assist you in selecting points in the drawing window you can setup the drawing aids displayed
in Figure 4-5 to suit the design requirements of your current project.

You can display a simple grid of points in the drawing window by selecting On from the Grid area
and entering suitable values in the X Spacing and Y Spacing fields, as shown in the following
figure:

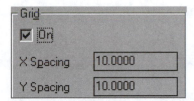

Figure 4-5: Settings for Grid

The grid which is displayed is for visual reference only. It is not considered part of your drawing,
nor is it plotted. To lock the points you select in the drawing window to a specific location, you
use Snap. This is an invisible grid of points used in conjunction with the grid settings.

Typically, you would select a value for snap which is half of the value entered for grid as shown
in the following figure:

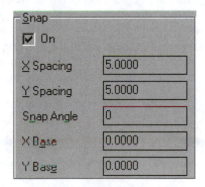

Figure 4-6: Settings for Snap

The settings shown in the previous figure will produce a rectangular grid with the Grid option set
to 10, and the Snap option set to 5. The grid display extends to the limits of your drawing.

Note: Setting the X value for grid or snap automatically sets the same Y value.

If you are working on a project that has objects placed at an angle, you can rotate the grid by
entering a value in the Snap Angle field. Entering a value of 45 would let you create the shape
shown in the following figure:

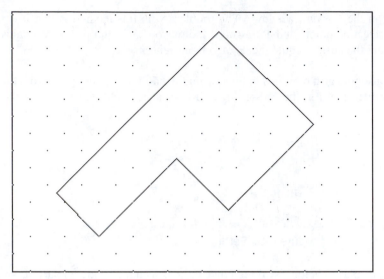

Figure 4-7: The Snap Angle set to 45

The X Base and Y Base fields let you set coordinate values for the base point of the grid. Typically these will be left at their default value of 0.0000.

In the Modes area of the Drawing Aids dialog box, the Ortho mode can be used to constrain the cursor movement to horizontal and vertical directions.

The Blips mode displays a temporary mark at the point in the drawing window which you select. The mark is in the shape of a plus sign as shown in the following figure:

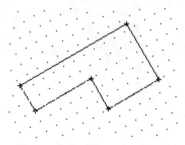

Figure 4-8: Blips displayed

The blips will remain in the drawing window until a REDRAW, REGEN, ZOOM, or PAN command is performed.

When you have selected the drawing aids you require, choose OK.

The grid Command

During your drawing session it may be necessary to change the current value of the grid spacing.

This could be done by accessing the Drawing Aids dialog box again, or by using the GRID command, which can only be accessed from the Command prompt.

The GRID command prompts for the current value and other options as follows:

```
Command: grid
Grid spacing(X) or ON/OFF/Snap/Aspect <10.0000>:
```

You can enter a new value for the grid spacing, or turn the grid display on or off.

To set the grid spacing to the current snap grid value, enter **Snap**. Different values for the X spacing and Y spacing can be set by entering **Aspect**.

The snap COMMAND

The snap grid spacing can be changed using the SNAP command which can only be accessed from the Command prompt.

The SNAP command prompts for the current value and other options, as follows:

```
Command: snap
Snap spacing or ON/OFF/Aspect/Rotate/Style <5.0000>:
```

You can enter a new value for the snap grid spacing, or turn snap on or off.

Different values for the X spacing and Y spacing can set by entering **a** at the Command prompt to select the Aspect option. The Rotate option lets you rotate the snap grid as shown in Figure 4-7.

The Style option sets either a Standard or Isometric snap grid. Standard is parallel to the XY plane of the current UCS, while Isometric displays a grid which is initially set to 30 and 150 degree angles.

Using the Status Bar

The Status Bar displays the coordinate location of your cursor and the current settings of Grid, Snap, Ortho, and other drawing aids, as shown in the following figure:

| 247.4874, 77.7817 ,0.0000 | SNAP | GRID | ORTHO | OSNAP | MODEL | TILE |

Figure 4-9: The Status bar

Using the Status bar you can toggle snap, grid, and ortho modes on or off by double-clicking the appropriate button.

Using Function Keys

You can also toggle the on or off status of Grid, Ortho, and Snap using the function keys. The function keys are mapped as follows:

▶   F7 - Grid

▶ F8 - Ortho

▶ F9 - Snap

Note: The previously modes mentioned can also be set to on or off using the following system variables:

▶ BLIPMODE - Controls whether marker blips are visible.

▶ GRIPMODE - Specifies whether the grid is turned on or off.

▶ SNAPMODE - Turns Snap mode on and off.

▶ ORTHOMODE - Constrains cursor movement to horizontal or vertical movements.

▶ FILLMODE - Specifies whether multilines, traces, solids, solid-fill hatches, and wide polylines are filled in.

Exercise 4-1: Establishing a drawing setup

A new AutoCAD drawing can be created using a number of techniques. In some cases a template file may exist which contains the settings you require for a specific project, while others may be created using setup options within AutoCAD.

In this exercise, you start a new drawing using Start from Scratch. You will then access the Units Control dialog box, determine the drawing limits, and use the Drawing Aids dialog box to create the initial drawing setup.

Using Start from Scratch

1. Start a new AutoCAD session.

2. From the Start Up dialog box, choose Start from Scratch.

Then select Metric as shown in the following figure:

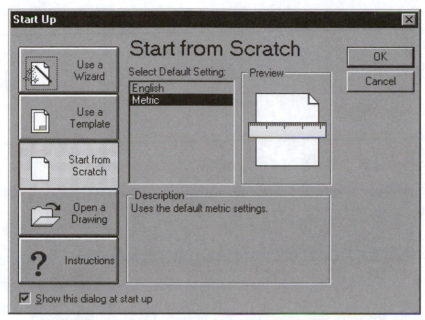

Figure 4-10: Selecting the Start from Scratch Metric option

3. Choose OK. A new drawing is created using the *acadiso.dwt* template file.

The drawing is currently named *drawing.dwg*. You will now save the drawing as *chap04.dwg*.

4. From the File menu, choose Save. The Save Drawing As dialog box is displayed. Make sure that the current directory is *c:\aotc\level1*.

5. Enter **chap04** in the File name field and then choose the Save button.

Using the Drawing Aids Dialog Box

The project requires an initial setup that will use decimal units with a precision of 2 decimal places. Angles will also be in a decimal format with a precision of 1 decimal place.

1. From the Format menu, choose Units.

2. Set Units and Angles as shown in the following figure:

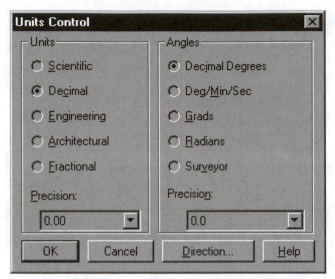

Figure 4-11: Setting Units and Angles

3. Choose the Direction button. The Direction Control dialog box is displayed.

4. You will use the following default values: Angle 0 Direction set to East, and Counter-Clockwise. Choose OK to close both dialog boxes.

Setting the Drawing Limits

The project you are working on is a drawing of a mechanical part which measures approximately 400 x 300 x 150 units. The part will be drawn at a scale of 1:2. You have to calculate the initial value for the drawing limits. This will include:

▶ Space for a front view, end view, and plan view

▶ Annotations around the drawings

The three views require a layout that is approximately 700 x 500. At a scale of 1:2 this will fit into a 420 x 297 sheet. Multiplying the sheet size by the scale gives you limits of 840 x 594. In this case a initial setting of 1000 x 800 will be appropriate.

1. From the Format menu, choose Drawing Limits. Accept the default value for the Lower left corner and enter **1000, 800** for the Upper right corner.

```
Command: '_limits

Reset Model space limits:

ON/OFF/<Lower left corner> <0.00,0.00>:

Upper right corner <420.00,297.00>: 1000,800
```

Note: These values can be changed at any time during the drawing session.

2.   From the View menu, choose Zoom. Then choose All to display the drawing limits.

Using the Drawing Aids Dialog Box

You will now setup the grid and snap for the initial drawing layout.

1.   From the Tools menu, choose Drawing Aids. The Drawing Aids dialog box will be displayed. Enter the values as shown in the following figure:

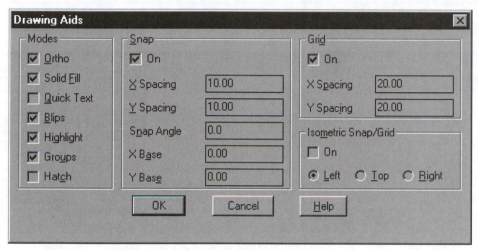

Figure 4-12: Setting the values in Drawing Aids

Then choose OK. Your drawing will display a grid that extends to the drawing limits and has a spacing of 20 units in the X and Y direction.

2.   Move the cursor around the drawing and note that the snap setting of 10 forces the cursor to snap to a grid with a spacing of 10 units in the X and Y direction.

3.   Toggle the grid and snap on and off by double-clicking GRID or SNAP on the Status bar. Note the changes caused by turning these values on and off.

4.   From the File menu, choose Save.

Conclusion

After completing this chapter, you have learned the following:

▶   Once you determine what the unit type and precision will be for a drawing, you can set these values using the Units Control dialog box.

▶   You can use the Drawing Aids dialog box to set grid and snap spacing and drawing modes such as Ortho and Blip.

▶   You determine the drawing scale factor by multiplying the length and the width of the plotting sheet size by the ratio at which you are scaling down the actual size of the drawing.

▶ You can use the LIMITS command to establish the coordinate limits of your drawing.

▶ The Status bar displays the coordinate location of your cursor and the current modes of Grid, Snap, Ortho, and other drawing aids.

Chapter 5

Creating Drawings Using Coordinate Entry

AutoCAD uses several point entry methods to locate points in a drawing plane. You can use several different types of coordinate and point entry methods to specify points. Each of these methods uses the Cartesian (rectangular) coordinate system. In this chapter, the following methods are explained:

- ▶ World Coordinate System (WCS) and UCS icon

- ▶ Absolute coordinate entry

- ▶ Relative coordinate entry

- ▶ Polar coordinate entry

- ▶ Direct distance entry

- ▶ Tracking

- ▶ X,Y,Z Point Filters

About This Chapter

In this chapter, you will do the following:

- ▶ Use coordinate entry to draw line segments.

The Cartesian Coordinate System

The *Cartesian coordinate system* has three axes X,Y, and Z that are used by AutoCAD to locate points and create objects. To simplify the discussion, this chapter focuses on the X and Y axes. The Cartesian coordinate system uses distances (in units) to locate points along intersecting axes, the horizontal "X" axes and the vertical "Y" axes. The intersection of these axes is called the origin, where X = 0 and Y = 0, denoted as (0,0). These axes divide the coordinate system into four quadrants each having positive, negative, or positive and negative X and Y values.

As you draw, you can enter a coordinate to locate a point. For example, you can draw a line by starting it at the 0,0 location in the coordinate system and ending it at the 3,4 location. Locations for the start (0,0) and end (3,4) point coordinate are shown in the following figure:

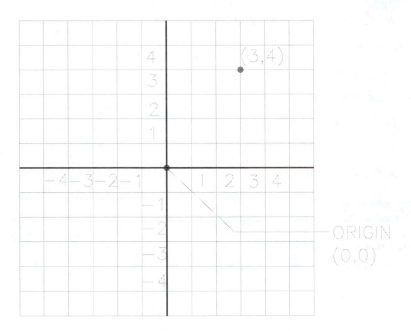

Figure 5-1: Two dimensional Cartesian coordinate system

World Coordinate System

When you start a new drawing session, by default you enter the World Coordinate System (WCS). This system consists of a horizontal X displacement, a vertical Y displacement, and a Z displacement that is perpendicular to the XY plane. The Z displacement is used for 3D drawing. All X,Y, and Z coordinate values are measured from the origin. The origin is located at the intersection of X, and Y axes (0,0). The origin is originally located in the lower left corner of a drawing. The WCS cannot be redefined, and all other user coordinate systems are based on the WCS.

The User Coordinate System (UCS) lets you establish your own coordinate origin. The UCS is movable, meaning that the origin can be moved to any desired orientation and its axes can be rotated. The User Coordinate System (UCS) icon is displayed by default in the lower left corner of the drawing window. The icon is used to help you better understand the location and orientation of the movable UCS. The X or Y arrows point in the positive direction of the axis. The W in the following figure indicates the WCS is current. The UCS icon is shown in the following figure:

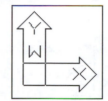

Figure 5-2: UCS icon

Methods for invoking the UCS command include:

▶ **Toolbar:** Standard

▶ **Menu:** Tools > UCS

▶ **Command:** UCS

Coordinate Entry

Absolute Cartesian Coordinate Input

Use the absolute coordinate system when you know the exact X and Y values of the point you want to place in the drawing window. The absolute coordinates method uses the Cartesian coordinate system to locate points in the drawing window. All points are measured from the 0,0 origin.

Relative Cartesian Coordinate Input

A relative coordinate is entered as the X and Y distance from the last point you specified. Use relative X,Y coordinates when you know the position of a point in relation to the previous point. For example, to locate a point in a relative direction of 4,5 from the first point specified, precede the next coordinate with the @ symbol. The following example demonstrates how to enter a relative coordinate:

At the Command prompt, enter **line**.
In response to the From point Command prompt, enter **0,0**.
In response to the To point Command prompt, enter **@4,5**.

Polar Input

A polar coordinate is entered as a relative distance and angle from an absolute coordinate or the last point specified.

To enter a polar coordinate, enter a distance and an angle, separated by an angle bracket (<). For example, to specify a point that is at a distance of 2.5 units from the previous point and at an angle of 45 degrees, enter @2.5<45.

By default, angles increase in the counterclockwise direction and decrease in the clockwise direction. To move clockwise, enter a negative value for the angle. For example, entering @3<-45 is the same as entering @3<315.

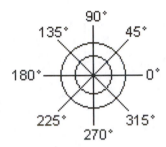

Figure 5-3: Polar coordinate angles

Direct Distance Entry

To enter points you can use a feature called direct distance entry. With direct distance entry, you can specify a relative coordinate by moving the cursor in the desired direction and then entering a distance.

Coordinate values are specified by moving the cursor to indicate a direction and then entering the distance from the first point. This is useful when you need to quickly specify a length. An efficient way to create object constrained to the current axes is to use direct distance entry combined with Ortho mode turned on.

Direct distance entry can be used with all commands except those that prompt you to enter real values. These commands include ARRAY, MEASURE, and DIVIDE.

Tracking

Tracking is a precision drawing tool that can be used to visually locate points relative to other points in your drawing. In addition to locating points, tracking is also used to find the center point of objects, to insert objects, and to place text at specified distances from known objects. Points are not placed in the display window until the tracking is turned off. To do this, press ENTER.

X,Y,Z Point Filters

Point filters let you specify one coordinate value at a time while ignoring other coordinate values. Used with object snaps, point filters can extract coordinate values from an existing object so you can locate another point.

Exercise 5-1: Combining Various Forms of Coordinate Entry

 AutoCAD provides various commands that you use to locate points in the drawing plane. In this exercise, you create an 8.5"x11" title sheet using the LINE command with absolute, relative, relative-polar, direct distance, and tracking coordinate entry methods.

Drawing the Title Block

1. Open the file *ms-85x11.dwg*. The drawing opens with Grid turned on and set to .25. The partially completed drawing looks like the following figure:

Figure 5-4: Ms-85x11.dwg

2. Now draw lines constrained to the X,Y axis by setting the Ortho mode on.

   Double-click the Ortho button in the Status bar to turn on Ortho mode.

3. First, draw the lines that make up the outline of the title block.

   From the Draw menu choose Line.

4. For the line starting point, use direct distance coordinate entry and Tracking mode combined with Ortho mode.

To identify which tracking direction to use, position the cursor in the top left corner of the drawing. In response to the LINE Command prompt, From point, enter **tk** to start Tracking mode and press ENTER.

5. For the first tracking point, enter **0,0** and press ENTER.

6. You are now in Tracking mode, which is like an invisible construction line mode. Using direct distance, you only need to position the cursor and enter a distance.

   In response to the LINE prompt, Next point, enter **1** and press ENTER. Notice that no line has been drawn but the tracking point has moved from the bottom left corner of the grid.

7. Press ENTER again to end Tracking mode and designate the line starting point.

8. Position the cursor on the right side of the window. In response to the LINE prompt, To point, enter **8** and press ENTER. Press ENTER again to end the LINE command. You have just drawn the first line of the title block.

9. Press ENTER to start the LINE command again. Position the cursor on the left side of the window and start Tracking mode by entering **tk** at the Command line and press ENTER.

10. For the first tracking point, enter **@** and press ENTER. The @ sign designates the last point entered.

11. In response to the LINE prompt, Next point, enter **1** and press ENTER. Press ENTER again to end Tracking mode and designate the starting point.

12. Position the cursor at the bottom of the window. In response to the LINE prompt, To point, enter **1** and press ENTER. You have just drawn the second line of the title block. Press ENTER again to exit the LINE command.

Drawing the Border

1. Now you draw the border boundary using the LINE command.

   From the Draw menu, choose Line.

2. For the line starting point, you use an absolute coordinate.

   In response to the LINE prompt, From point, enter **0,0** and press ENTER.

3. For the next point, you use another absolute coordinate specifying an X value of 8 and a Y value of 0.

   In response to the LINE prompt, To point, enter **8,0** and press ENTER.

4. For the next point, you use a relative polar coordinate 10.5 units at an angle of 90 from the last point entered.

   In response to the LINE prompt, To point, enter **@10.5<90**, and press ENTER.

5. For the next point, you use a relative coordinate with a negative X value of minus 8 and

a Y value of 0 from the last point entered.

In response to the LINE prompt, To point, enter **@-8,0** and press ENTER.

6.   For the last point of the border you use the LINE command option, Close, to complete the border rectangle.

In response to the LINE prompt, To point, enter **c** for close and press ENTER.

7.   This exercise is complete. From the File menu, choose Save to save the drawing.

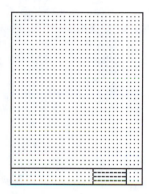

Figure 5-5: Completed ms-85x11.dwg

Conclusion

After completing this chapter, you have learned the following:

▶   You use various coordinate entry methods to create objects.

Chapter 6

Using Layers

In this chapter, you learn how to use layers to control the color, linetype, and visibility of objects in your drawing.

About This Chapter

In this chapter, you will do the following:

▶ Use the Layer Dialog Box to create multiple layers.

▶ Control multiple layers with the -LAYER Command.

▶ Make a layer current by selecting an object.

▶ Load linetypes and apply linetype scaling.

Layer Standards

Layering standards provide consistent naming schemes for drawing layers that will be used in all of your AutoCAD drawings to manage objects. The use of layering standards for simple and complex drawings is important whether you work alone or in a large organization. Layering standards facilitate your control of objects in the AutoCAD environment for yourself and anyone else who works with your AutoCAD files.

Layering standards are application-specific by industry, with only a limited number of organizations publishing standards at the national or international level. Developing your own standards will let you be more consistent and efficient, even when formal layer control is not required. Before you create your own standards, you may want to investigate existing standards for use as is, or modified to suit your organization's requirements.

The use of layering standards facilitates control of the following:

- Consistency in drawing practices and locating objects

- Various linetypes

- Drawing organization

- Streamlined display and overlay of objects

- Simplifying selection sets for various commands

- Protection of objects when you freeze or lock layers

- Control of object screen colors for clarity

- Controlling pen widths for plotting

- Preventing objects from plotting

- Layer management of external references

- Compatibility with custom automation (*.lsp* files, scripts, and Active X)

- Assign rendering materials

Tips for Layering Standards

It is impossible to cover every application of layering standards in this course. Instead, use the following tips as a guide for creating your own standards:

- Use names that adequately describe layer contents, but are not so long that they will be truncated in the Layer Control window of the Object Properties toolbar.

- Use a prefix for dialog box and Layer Control window sorting.

- Use a suffix to control groups of layers with wildcard symbols in local or externally referenced files. Enter **-la** at the Command prompt to control multiple layers at the Command prompt, and use wildcards with the Command prompt options.

Using the Layer Dialog Box

By creating objects on layers, objects can be grouped in sets. You can then define the color, linetype, and visibility of the objects in these sets. These layers may be turned on and off for display or performance purposes. This section covers the different methods for managing layers in your drawing.

> Note: Objects placed on the DEFPOINTS layer will be displayed in the Drawing window, but will never plot.

Layer Tab

The Layer Tab of the Layer & Linetype Properties dialog box has the most features for managing the layers in your drawing. Using the Layer tab, you can: add new layers to your drawing, make a layer current, rename an existing layer, and delete layers from the list. You can turn layers on and off, freeze and thaw layers globally or by viewport, and lock and unlock layers.

Methods for opening the Layer & Linetype Properties dialog box include:

- **Toolbar:** Object Properties

- **Menu:** Format > Layer

- **Command:** LAYER

The Layer & Linetype Properties dialog box is displayed, as shown in the following figure:

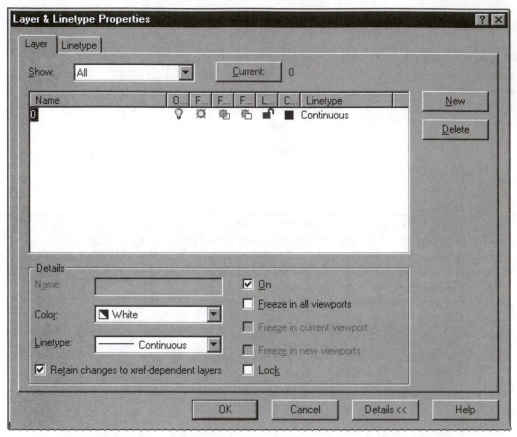

Figure 6-1: The Layer tab of the Layer & Linetype Properties dialog box

The Layer Tab Status bar displays a row of options used to set layer properties. After each layer is created a row is formed in the dialog box that displays each option setting. Layer properties are modified by clicking the associated icon on or off. A description of the items located on this bar follows:

▶ *Name* - The Name option displays layer names and lets you rename layers.

▶ *On* - The On option turns layers on and off. Layers that are turned on are displayed and plotted. Layers that are turned off are not visible, printed, or plotted. However they are still included in regeneration calculations.

Freezing and Thawing: When a layer is frozen it is not visible, plotted, or calculated by the computer when a regeneration is performed. When a layer is thawed it is visible, plotted, and included in the regeneration of a drawing.

▶ *Freeze in All Viewport* - Freezes all the selected layer viewports.

▶ *Freeze in Current Viewport* - Freezes all selected layers in the current floating viewport.

▶ *Freeze in New Viewport* - Freezes selected layers in new floating viewports.

▶ *Lock* - This option determines if a layer is locked or unlocked. Locked layers cannot be edited, unlocked layers can be edited.

▶ *Color* - This option lets you change or set the layer color by selecting a color from the Select Color dialog box.

▶ *Linetype* - The linetype option is used to change the layer linetype of one layer or a group of layers.

Using the Show drop-down list, you can determine which types of layers are displayed in the layer name list. You can display layer names based on whether the layers are referenced in an external drawing, or whether the layers contain objects. You can also filter layers based on their name, state, color, linetype, or whether they are frozen in the current viewport or in new viewports.

Choosing Set Filters dialog in the Show drop-down list displays the Set Layer Filters dialog box, as shown in the following figure:

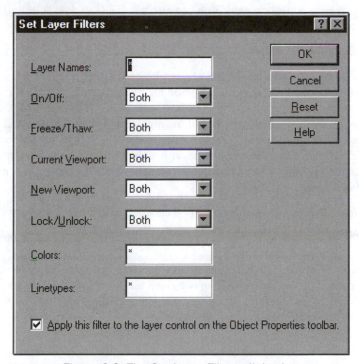

Figure 6-2: The Set Layer Filters dialog box

Use the Set Layer Filters dialog to show layers according to very specific criteria, such as all the frozen layers with blue dashed linetypes. When you choose OK to return to the Layer & Linetype Properties dialog box, only those layers that meet the criteria will be listed.

If you want to sort the order in which layers are displayed in the list box, choose the column headings. The first selection of the Name column puts the layers in descending order (Z to A, then numbers), and a second selection puts the layers in ascending order (numbers, A to Z). Choosing the status column headers such as Frozen or Locked initially puts the frozen or locked layers first in the list. When you choose the color or linetype columns, the color column is in AutoCAD Color Index (ACI) order, while the linetype column is displayed in alphabetical order. To change the status of a property for a layer, choose that property from its name row. If you choose Color or Linetype, the Set Color or Select Linetype dialog box is displayed. You can highlight more than one layer with the SHIFT or CTRL keys in combination with the left mouse button, similar to Windows file selection options. If more than one layer is highlighted, and you choose one of the properties, that property is changed for all the highlighted layers. If one layer or more is highlighted and you choose the property of another layer, the highlighting is canceled and the change is made for the chosen layer only.

Descriptions of the remaining dialog box options are as follows:

▶ *Select All and Clear All* - To use these options, right-click your mouse anywhere in the dialog box. The Select All and Clear All options are displayed as a cursor menu. You can also use CTRL + A to select all the layers.

▶ *Current* - Makes a single highlighted layer current.

▶ *New* - Creates a new layer. If no layers are highlighted, the new name is Layer1 and the properties are based on the properties of Layer0. If Layer1 is already present, the new layer is Layer2, and so on. If a layer is highlighted, the new layer gets the properties of the highlighted layer. Pressing ENTER twice creates new layers. You can enter multiple new layer names separated by a comma. Regardless of how layer names are typed, all layers names are formatted starting with an uppercase character while the remaining characters are lowercase.

▶ *Delete* - Deletes selected layers. You cannot delete the following layers: Layer0, DEFPOINTS, the current layer, layers containing objects, or external reference-dependent layers.

▶ *Details* - Lets you switch between an enlarged view of the Layer tab or a condensed view. This switching is termed unfolding (when enlarging the dialog box) and folding. The next time you open the dialog box, it is displayed in the state in which it was left.

-LAYER Command

Use the -LAYER command to create new layers, set the current layer, and set the color and linetype for designated layers right from the Command prompt. With this command, you can also turn layers on and off, lock or unlock layers, freeze or thaw layers, and list defined layers.

Entering -LAYER at the Command prompt displays the following list of options:

?/Make/Set/New /ON/OFF/Color/Ltype/Freeze/Thaw /LOck/Unlock:

Layer Drop-down List in the Object Properties Toolbar

Using the Object Properties toolbar, you can directly edit the properties of gripped objects when the select mode, Noun/Verb Selection is enabled. *Gripped objects* are objects selected without an active command and are displayed highlighted with boxes at control points along the objects.

With the Layer drop-down list, you can change the on/off, unlocked/locked, thawed/frozen status, and current layer status of layers. In the Layer Drop-down list you can select icons multiple times to control layer states. You cannot, however, change a layer's color. You can use the Layer drop-down list to change the layers of multiple objects in a manner similar to the DDCHPROP and CHPROP commands.

Making a Layer Current by Selecting an Object

The Make Object Layers Current command sets the current layer to match that of a selected object. This is a faster, more accurate, and simpler method for changing layers than using the Layer & Linetype Properties dialog box. This command is particularly useful for complex drawings or for making a dimension or annotation layer current.

To use this command, on the Object Properties toolbar, choose the Make Layer Current button. At the `Select Object` prompt, select an object whose layer you want to make current.

Exercise 6-1: Working with Layers

In this exercise, you use the Layer & Linetype Properties dialog box to create new layers and assign them different colors. You then set the new layer to current. Observe that objects you draw display the color properties of the current layer. Next, at the Command prompt use the -LAYER command to control the visibility of multiple layers. To conclude this exercise, you learn different methods for making a layer current.

Defining Layers

1. Open the file, *wallset.dwg* drawing. The drawing looks like the following figure:

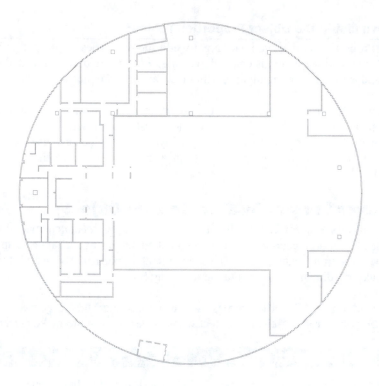

Figure 6-3: Wallset.dwg

2.   From the Format menu, choose Layer. The Layer & Linetype Properties dialog box is displayed, as shown in the following figure:

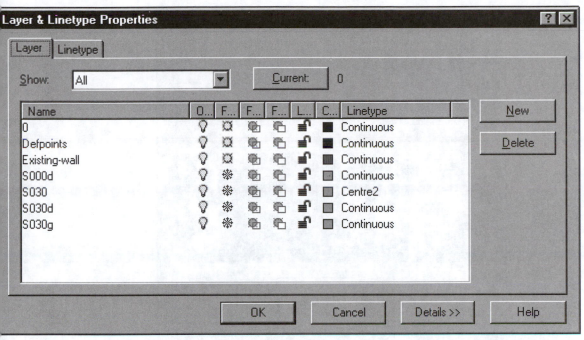

Figure 6-4: The Layer & Linetype Properties dialog box

3.   You now create a new layer.

In the Layer & Linetype Properties dialog box, choose the New button. Under the Name column heading, notice that a new layer named LAYER1 has been created.   Enter **border** and then press ENTER. The layer named BORDER has now replaced the layer named LAYER1.

4.   Now, assign a red color to the BORDER layer.

While BORDER is still highlighted, under the Color column heading, choose the square in the BORDER layer row. The Select Color dialog box is displayed, as shown in the following figure:

Note: To expand a column heading, move the cursor into the column heading you want to expand. Notice that the cursor changes into a splitter bar. Drag the splitter bar horizontally until the column heading displays the full description.

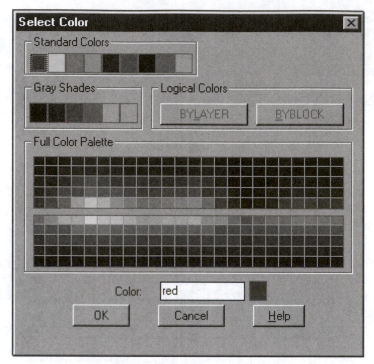

Figure 6-5: Select Color dialog box

5.   In the Select Color dialog box, in the Standard Colors area, select the red square. Notice that at the bottom of the dialog box, the color description is displayed along with a solid filled square that shows the selected color. Choose OK.

6.   You now create multiple new layers. To do this, separate the layer names with commas.

       Choose the New button and enter **demo-wall, new-wall**. Press ENTER. You have just created two wall layers.

7.   Choose OK to close the Layer & Linetype Properties dialog box.

Working with the Current Layer

1.   Press ENTER to repeat the last command. The Layer & Linetype Properties dialog box is displayed.

2.   Select the Border layer and choose the Current button. Border is now the current layer and its name is displayed next to the Current button. Choose OK to close the Layer & Linetype Properties dialog box.

3.   Using the PLINE command, a boundary will now be placed around the building using absolute, relative, polar, and direct distance coordinate entry methods.

       At the Command prompt, enter **pline**, or its keyboard alias, **pl**. Enter an absolute coordinate starting point of **0,0** and press ENTER.

4. For the second endpoint, enter a relative coordinate of **@200000,0** and press ENTER.

5. For the third endpoint, enter a relative polar coordinate of **@110000<90** and press ENTER.

6. For this next endpoint, use the direct entry method. Be sure that Ortho mode is on by pressing the F8 key and observing <ortho on> at the Command prompt or the highlighted ORTHO button in the Status Bar at the bottom of the screen.

   For the fourth point, move the mouse cursor to the upper left corner of the drawing window, enter **200000** and press ENTER.

7. To complete the boundary use the PLINE command option Close.

   Enter **c** and press ENTER. Since the Polyline was created on the layer border it displays the color properties of that layer.

Command prompt Layer Control

1. When working with similarly named layers it is sometimes more efficient to use the LAYER command at the Command prompt than the dialog box. You now use the Command prompt to globally control the visibility of multiple layers.

   At the Command prompt enter **-layer** or its keyboard alias **-la**.

2. You use the -LAYER command option, Thaw, to change the visibility of all structural layers.

   Enter **thaw** or the letter **t**. In response to Layer names to thaw, enter **S*** and press ENTER twice. This thaws all layers that begin with S.

   ```
   Command: -la

   -LAYER ?/Make/Set/New/ON/OFF/Color/Ltype/Freeze/Thaw/LOck/Unlock:
   t

   Layer name(s) to Thaw <>: s*

   ?/Make/Set/New/ON/OFF/Color/Ltype/Freeze/Thaw/LOck/Unlock:
   ```

   Press ENTER to complete the command.

3. All layers that begin with S are now displayed in the drawing window.

4. Enter **la** at the Command prompt. In the Layer tab of the Layer & Linetype Properties dialog box, observe the thawed status of all the layers that begin with S. Choose OK.

Making a Layer Current

In the first part of this exercise, "Working with the Current Layer", you learn how to set a layer current from the Layer dialog box by selecting the current button. You can also set an existing layer current by selecting the Layer Control drop-down list from the Object Properties toolbar.

1. Select the EXISTING-WALL layer from the Layer Control drop-down list in the Object

Properties toolbar. The EXISTING-WALL layer is now the current layer, displayed in the layer control drop-down list, as shown in the following figure:

Figure 6-6: Object Properties Toolbar with the Layer Control selected

3. Choose the Make Object's Layer Current button in the Object Properties Toolbar.

4. Select one of the center lines that represents the column grid. The layer named S030 is now displayed as the current layer in the Layer Control drop-down list, as shown in the following figure:

Figure 6-7: Object Properties Toolbar with Make Object's Layer Current button selected

5. Save your drawing as *wallset1.dwg* using the SAVEAS command.

Conclusion

After completing this chapter, you have learned the following:

▶ Using the Layer tab of the Layer & Linetype Properties dialog box, you can: change a layer's status, make a layer current, add new layers to the layer name list, rename an existing layer, assign color and linetype to the layers, and delete layers from the list.

▶ Use the Set Layer Filters dialog box to show layers according to very specific criteria, such as all frozen layers with blue dashed linetypes.

▶ Use the -LAYER command to create new layers, set the current layer, and set the color and linetype for designated layers from the Command prompt.

▶ With the Layer drop-down list, you can change the on/off, unlocked/locked, and thawed/frozen status, and set the current layer.

▶ The Make Object Layers Current command sets the current layer to match that of a selected object.

▶ To change the color of an object, you select the object and then choose a different color from the Color Control drop-down list.

▶ If the current color is set to ByLayer, objects are created in the color assigned to the current layer.

▶ The Property Painter prompts you to copy properties from one object to another. The basic properties which can be copied from the source object to the destination object are: color, layer, linetype, linetype scale, and thickness.

Chapter 7

Display Commands

In this chapter, you learn how to use display commands and control the view of objects displayed in the drawing window.

About This Chapter

In this chapter, you will do the following:

▶ Use the various ZOOM command options.

▶ Adjust the display using Realtime ZOOM and PAN modes.

▶ Use the Aerial View feature to zoom and pan different parts of a drawing.

▶ Use the ZOOM command while in an active command.

▶ Create and restore views with the View Control dialog box.

▶ Use tiled viewports to display different parts of a drawing.

▶ Learn when to use the REDRAW, REGEN, and REDRAWALL commands and experience the performance difference.

Controlling the Display

Display commands provide you with different options for changing the view of your drawing while it is being created. These commands make it easier to work with your drawing, and see the overall effects of changes. When you use drawing or editing commands, you can utilize the transparent PAN command and ZOOM command options to change the view and magnification of a drawing. You can also use display commands for saving and restoring specific views, or to display several views created from tiled viewports.

Using the ZOOM Command

When you create a drawing, you may need to change the way objects are viewed in the drawing window. To achieve this, the orientation, magnification, or position of the drawing may have to be adjusted. The ZOOM command options let you change the view by increasing or decreasing the size of displayed images. You zoom in to magnify objects so you can see more details. You zoom out to reduce objects in the drawing window and view a larger portion of the drawing.

Zooming does not change the true size of a drawing or object. It only changes the size of the view in your drawing window. You can view the entire drawing, specify a display window, or zoom to a specific scale.

Window

The Zoom Window option lets you zoom in on an area of your drawing by using a window to specify the viewing boundaries. When you enter the Zoom Window option, AutoCAD displays a rectangular window in the drawing window. Select a point to specify where you want your view to start, then move the window to cover the area or section you want to magnify.

Methods for invoking the zoom window option include:

> ▶ **Toolbar:** Standard > Flyout, Zoom

> ▶ **Menu:** View > Zoom > Window

> ▶ **Command:** ZOOM > Window

Previous

The Zoom Previous option displays the last view of your drawing. This option lets you restore as many as ten previous views.

Methods for invoking the Zoom Previous option include:

> ▶ **Toolbar:** Standard

> ▶ **Menu:** View > Zoom > Previous

> ▶ **Command:** ZOOM > Previous

All

The Zoom All option lets you view the entire drawing in the current viewport. The display shows the drawing limits and all objects even if the objects extend outside of the drawing limits.

Methods for invoking the Zoom All option include:

▶ **Toolbar:** Standard > Flyout, Zoom

▶ **Menu:** View > Zoom > All

▶ **Command:** ZOOM > All

Extents

The Zoom Extents option displays the region of the drawing where all objects you draw are located. The display is based just on drawing objects, the drawing limits are not considered to recalculate the display.

Methods for invoking the Zoom Extents option include:

▶ **Toolbar:** Standard > Flyout, Zoom

▶ **Menu:** View > Zoom > Extents

▶ **Command:** ZOOM > Extents

Center

The Zoom Center option lets you change the displayed size of an object and locate it in the center of the viewport. When you enter the Zoom Center option, you are prompted to locate a center point on the drawing plane. The next command prompt lets you enter a magnification value or a height. The magnification value is followed by an **x** which is the relative magnification factor. For the Height option, enter smaller numbers for the height to enlarge the image size, and enter larger numbers for the height to decrease the size.

Methods for invoking the Zoom Center option include:

▶ **Toolbar:** Standard > Flyout, Zoom

▶ **Menu:** View > Zoom > Center

▶ **Command:** ZOOM > Center

Dynamic

The Zoom Dynamic option uses a viewbox to adjust the display. The view box represents your viewport. This means that the region of your drawing that you select with the view box will be displayed in the entire drawing window. You can enlarge, reduce, and move the viewbox around your drawing to select the desired view.

Methods for invoking the Zoom Dynamic option include:

- ▶ Toolbar: Standard > Flyout, Zoom
- ▶ Menu: View > Zoom > Dynamic
- ▶ Command: ZOOM > Dynamic

Scale

The Zoom Scale option uses a scale factor to adjust the display. The value you enter should correspond to the limits of the drawing.

The Scale (X) option requires you to use positive numbers to change the magnification factor. When you enter the **X** option, the current display can be enlarged or reduced by a certain multiple. For example, entering **2** at the Zoom Scale Command prompt will display your drawing at twice its size, relative to its full view. However, entering **2x** at the Zoom Scale Command prompt will display your drawing at twice the size as displayed in the current view.

You use the Scale (XP) option for scaling the display in floating viewports created in paper space. This is useful when you are plotting layouts of scaled multiview drawings.

Methods for invoking the Zoom Scale option include:

- ▶ **Toolbar:** Standard > Flyout, Zoom
- ▶ **Menu:** View > Zoom > Scale
- ▶ **Command:** ZOOM > Scale

Realtime

When working with complex drawings, you spend significant time using the ZOOM and PAN commands. After you invoke the command, the Realtime Zoom cursor is displayed, drag the cursor down to decrease or up to increase the image. To activate the Realtime ZOOM / PAN menu right-click the mouse.

The ZOOM/PAN cursor menu options include:

- ▶ *Exit* - cancels the Realtime option and returns you to the Command prompt
- ▶ *Pan* - Switches from Realtime ZOOM to PAN
- ▶ *Zoom* - Switches from PAN to Realtime ZOOM
- ▶ *Zoom window* - Displays a specified window and returns to REALTIME ZOOM
- ▶ *Zoom Previous* - Restores the previous view and returns to the Realtime option

▶ *Zoom Extents* - Displays the drawing extents and returns to the Realtime option

Methods for invoking the Realtime ZOOM command include:

▶ **Toolbar:** Standard

▶ **Menu:** View > Zoom > Realtime

▶ **Commands:** ZOOM REALTIME

The Zoom/Pan Cursor menu is shown in the following figure:

Figure 7-1: The ZOOM/PAN cursor menu

Pan

With the PAN command, you can move the drawing around the drawing window without changing the magnification. The two options that let you change the display are Realtime and Point.

Methods for invoking the PAN command include:

▶ **Toolbar:** Standard

▶ **Menu:** View > Pan > Realtime

▶ **Command:** PAN

When you select the Realtime PAN option, the cursor changes into a hand cursor. To change the location of your drawing, press the left mouse button. This locks the cursor into its current location relative to the coordinate system used in the current viewport. As you move the mouse, the drawing image pans to a new location. The view of the objects in the drawing window shift in the same direction as the cursor.

The PAN Point option lets you specify a single point. AutoCAD then uses the selected point and the current view orientation to displace the image. You also can specify two points, AutoCAD computes the displacement between the two points then moves the image in the drawing window. You can also pan by entering absolute or relative coordinates at the Displacement prompt. To invoke the Pan Point option, select Pan from the View menu, then choose Point.

The PAN command also has preset options that move the drawing in the specified direction. You can access these options from the View menu. They include Left, Right, Up, and Down. A view of the Pan cascading menu is shown in the following figure:

Figure 7-2: Pan cascading menu options

Exercise 7-1: Controlling the Display

 You often use the ZOOM command options to change the view in the drawing window. In this exercise, you use the ZOOM options to display new drawing views.

Using the Zoom Command

1. Open the file *display.dwg.* The drawing looks like the following figure:

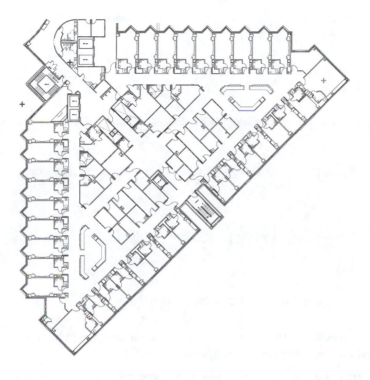

Figure 7-3: Display.dwg

2. From the View menu, choose Zoom, then choose Window.

3. In response to the `First corner:` prompt, select the magenta cross below and to the left of the square stairwell.

4. For the `Other corner:`, select the magenta cross above and to the right of the square stairwell. The following figure shows the result of using zoom window:

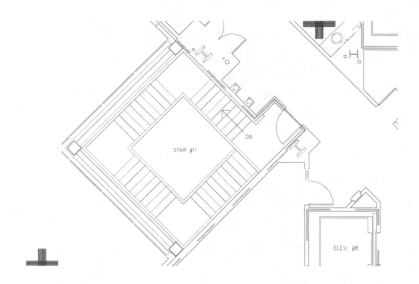

Figure 7-4: Result of using ZOOM Window

5. To recall a previous display, you can use the Zoom command option Previous. At the Command prompt, enter the command alias **z**.

Enter **p** at the Command prompt. The previous display is now displayed again.

6. To display all visible objects in the whole drawing, you can use the Zoom Extents option.

From the View menu, choose Zoom, then choose Extents. All visible objects in the drawing are now displayed.

7. To center an object in the middle of the drawing window, use the Zoom Center option.

At the Command prompt, enter **z**. Then enter **c** for the center option.

8. Select the green cross located in an office near the upper right corner of the building.

9. When prompted for Magnification or Height, enter **1500,** then press ENTER. The office with the green cross is now centered in the drawing window.

Using Realtime Zoom and Pan

1. The REALTIME PAN and ZOOM commands let you quickly obtain the display you want.

To use the REALTIME ZOOM command, enter **z** and accept the default command option <realtime> by pressing ENTER.

2. The Realtime Zoom cursor is now displayed as a magnifying glass with plus and minus signs. Press the left mouse and drag the Realtime Zoom cursor up and down the drawing window. When the desired view is displayed in the drawing window, release the left mouse button.

3. Press the right mouse button. The Pan/Zoom cursor menu is displayed. Select the Pan option, as shown in the following figure:

Figure 7-5: Realtime Pan and Zoom cursor menu

4. The Realtime Pan cursor is now displayed as a hand in the drawing window. Position the Realtime Pan cursor over the office building. Drag the cursor around the drawing window, and when the desired view is displayed in the drawing window, release the left mouse button and press ENTER.

5. This concludes the ZOOM Command exercise. You are encouraged to explore the other Zoom command options and command entry methods. Do not save your drawing.

The Aerial View Feature

The Aerial View feature is a viewing aid that saves drawing time by displaying your entire drawing in a separate window. You can then use the ZOOM and PAN command options within the window to change the view of your drawing on the drawing window. If you keep the Aerial View window open while you work, you can pan and zoom around your drawing without entering commands at the Command prompt, or accessing them from the View menu.

Methods for invoking the Aerial View window include:

▶ Toolbar: Standard

▶ Menu: View > Aerial View

▶ Command: DSVIEWER

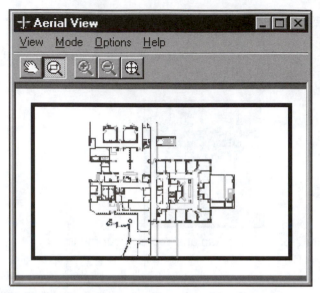

Figure 7-6: The Aerial View window

When you enter the Aerial View window, AutoCAD displays the Aerial View window by default, in the same location and with the same size as it was located in the last drawing session. This means that you may have to move or change the size of the Aerial View window so that you can have a better view of the display window. To do so, use the left mouse button to select the title bar. Drag the window to a desired location and release the left mouse button. To change the size of the window, move your cursor to one of the four corners. When you see the doubled-pointed arrow, drag the cursor to change the window size.

The Aerial View window contains the following four menus labeled View, Mode, Options, and Help.

The View menu lets you select three zoom options to change the magnification of the Aerial View. These include Zoom In, Zoom Out, and Global.

> ▶ *Zoom In* - Increases the magnification of the drawing in the Aerial View by zooming in by a factor of 2, centered on the current view box.

> ▶ *Zoom Out* - Decreases the magnification of the drawing in the Aerial View by zooming out by a factor of 2, centered on the current view box.

> ▶ *Global* - Displays the entire drawing and the current view in the Aerial View window.

When you display the entire drawing, the Zoom Out menu item and button are shaded out and become inactive. When the view fills the Aerial View window, the Zoom In menu item and icon become inactive. At times both of these conditions may exist, especially after zooming to the drawing's extents.

The Mode menu activates the Pan mode or the Zoom mode. When you choose the Pan menu item you will notice that the Pan button is activated. Pan lets you move the view box around the drawing without changing its size. Selecting Zoom activates the zoom window button. The zoom window button changes the view by increasing or decreasing the view box size.

The Options menu provides the following toggles for automatic viewport display and dynamic updating of the drawing:

> *Auto Viewport* - When Auto Viewport is selected, it automatically displays the active Model space viewport. When it is turned off, the Aerial view is not updated to match the active viewport.

> *Dynamic Update* - Controls whether or not AutoCAD updates the Aerial View while you edit the drawing.

The Help menu contains one option, Aerial View Help. Choose this item to open the main AutoCAD help menu which will display information on the Aerial View.

Using Named Views

When you are working on large drawings with many areas or details, you may find it time consuming to continuously use the ZOOM and PAN commands. To save time, create named views of areas in your drawing that you will use often. You create named views by using the View Control dialog box and the VIEW command. A view can be an enlarged section of a drawing that shows details, or a portion of a drawing, such as the lower-right quadrant. Once you create a view, it can be displayed at any given time.

View Control Dialog Box

You can create named views by using the options in the View Control dialog box.

Methods for invoking the View Control dialog box include:

> **Toolbar:** Viewpoint or Standard

> **Menu:** View > Named Views

> **Command:** DDVIEW

The View Control dialog box is displayed, as shown in the following figure:

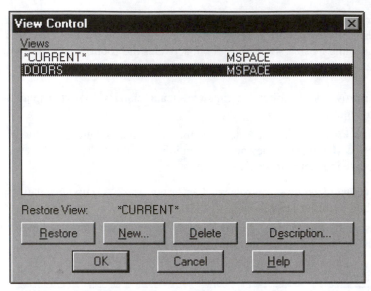

Figure 7-7: The View Control dialog box

A brief description of the options located in the View Control dialog box follows:

The Views list box lists all the drawings that you have named along with the default view *CURRENT*. If a named view was created in model space, MSPACE follows the name. If it was created in paper space, PSPACE follows the name.

The Restore button restores any of the listed views. To restore views, select the view name from the Views list box, then choose the Restore button. The name of the view to be restored is then displayed at the Restore View label. To display your selected view, choose OK.

The New button displays the Define New View dialog box. The New option lets you create new views. A brief description of the items located in this dialog box are as follows:

▶ *New Name* - Lets you specify the name of the view.

▶ *Current Display* - Uses the current display as the new view.

▶ *Define Window* - Uses the `First Corner` and `OtherCorner` **X,Y** coordinate values to define the location of the view window. Selecting Define Window also activates the Window button so that you can define a view by using the pointing device to specify points.

▶ *Window* - Defines a view window using points specified by the pointing device. The coordinates of the window are displayed under First Corner and Other Corner.

▶ *Save View* - Saves the new view and returns to the View Control dialog box where the new view is listed.

The Define New View dialog box is shown in the following figure:

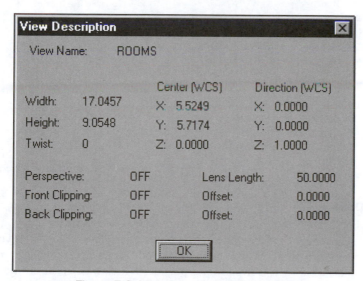

Figure 7-8: The Define New View dialog box

If you want to remove a named view from the dialog box, select the view from the Views list box, then choose the Delete button. The Delete button is activated when you select a named view. (If a view is not selected the button is not available.)

When you select the Description button, a dialog box with details of the view is displayed. An example of the View Description dialog box is shown in the following figure:

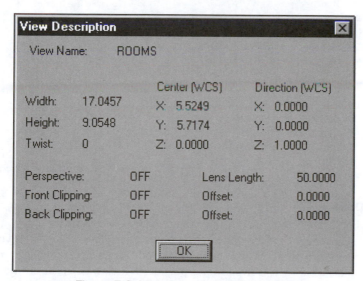

Figure 7-9: View Description dialog box

View Command

You use the VIEW command to create and restore named views. To created a named view, enter **view** at the Command prompt. Using the Command prompt version of the VIEW command does not display the View Control dialog box. You must enter all information about the named view at the Command prompt. The following options are displayed at the Command prompt:

- ▶ *?* - Lists the named views in the drawing and illustrates whether they are in model space or paper space.

- ▶ *Delete* - Removes one or more named views.

- ▶ *Restore* - Restores the named view to the display screen.

- ▶ *Save* - Saves the current display as a named view.

- ▶ *Window* - Saves a specified windowed area as a named view.

Using Display Commands Transparently

When the ZOOM, PAN, Aerial View, VIEW, and REDRAW commands are entered at the Command prompt preceded by an apostrophe, they become transparent display commands. This means that they can be accessed and used while you are operating another command. When you exit the transparent command, the command in which you were previously working becomes active. All of the commands discussed in this chapter are transparent commands except for REGEN, VIEWRES, and VPORTS.

Examples of the view and pan commands being used transparently are displayed in bold text in the following line:

```
Command: line
LINE From point: 'view
>>?/Delete/Restore/Save/Window: r
>> View name to restore: foyer
Resuming LINE command.
From point:
To point: 'pan
>>Press Esc or Enter to exit, or right-click to activate
pop-up menu.
Resuming LINE command.
To point:
```

Tiled Viewports

A new drawing session begins by default, using a single viewport that fills the drawing window. A *viewport* is a rectangular region of the drawing window that displays a portion of your drawing in model space. Model space viewports can be used for both 2D and 3D drawings. You can use the PAN or ZOOM commands in each viewport to display desired views of your drawing. You can divide the drawing window to show additional viewports which display different parts of your drawing. These regions are called tiled viewports.

You create Multiple viewports using the VPORTS command. You can create as many as 48 tiled viewports in model space that can be viewed at one time. The VPORTS command options let you vary the arrangement of the viewports. Viewports are created by entering information at the Command prompt, or by using the Tile Viewport Layout dialog box.

Note: To see the viewports, the TILEMODE system variable must be set to 1.

Methods for invoking the VPORTS command include:

▶ **Menu:** View > Tiled Viewports

▶ **Command:** VPORTS

The Tile Viewport Layout dialog box, as shown in the following figure:

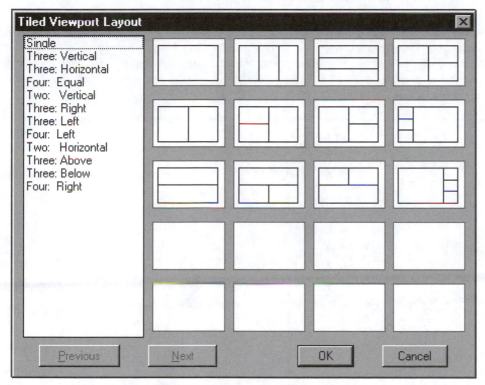

Figure 7-10: The Tiled Viewport Layout dialog box

You can choose the number and layout of your viewports by selecting an option from the Tiled Viewport layout dialog box and choosing OK. When you enter VPORTS at the Command prompt, you also have a list of options that include:

▶ *Save* - Saves the current viewport configuration using a specified name.

▶ *Restore* - Restores a previously saved viewport configuration.

▶ *Delete* - Deletes named viewport configurations.

▶ *Join* - Combines two adjacent viewports into one large viewport. The new viewport inherits the view of the current viewport, or a selected viewport which will be considered the dominant viewport.

▶ *Single* - Creates a single viewport, using the view from the current viewport.

▶ *?* - Displays the identification numbers and screen positions of the active and previously saved viewport configuration.

▶ *2* - Divides the drawing window into two viewports.

▶ *3* - Divides the drawing window into three viewports.

▶ *4* - Divides the drawing window into four equal viewports.

Although several tiled viewports may be active (displayed in the drawing window), they can only be current one at a time. When you are working in model space the cursor is displayed only in the current viewport. The current viewport is shown with a heavy border and displays the drawing cursor. You can only select objects or enter points from the current viewport. To make a viewport current, move the pointing device to the viewport and press the left mouse button. A current viewport is shown in the following figure:

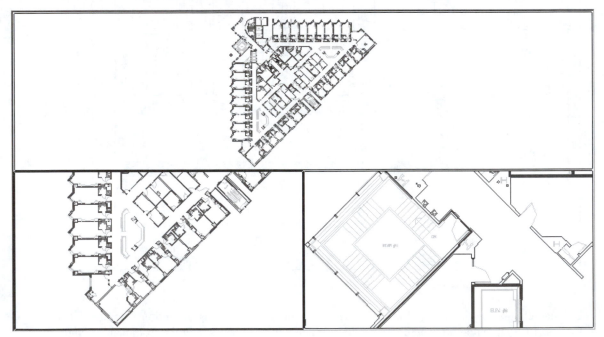

Figure 7-11: Tiled viewports with the bottom left viewport current

Exercise 7-2: Working with Views and Viewports

 In this exercise, you create named views, tiled viewports, and use the Aerial View to change the view in the drawing window.

Using Views

1. Open the file *display.dwg*. The drawing looks like the following figure:

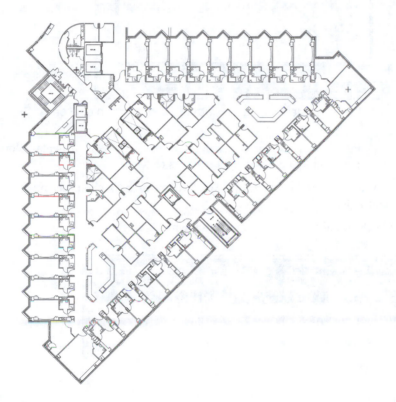

Figure 7-12: Display.dwg

2. From the View menu, choose Aerial View. The Aerial View window is displayed.

3. In the Aerial View window, choose the Zoom button, as shown in the following figure:

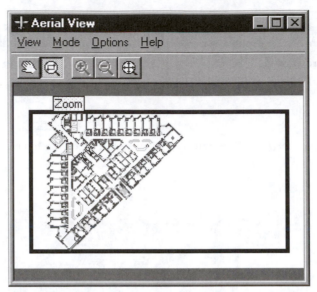

Figure 7-13: The Aerial View dialog box with the Zoom button chosen

4. Position the Aerial View cursor crosshairs over the lower left magenta-colored cross. Drag the rectangle out to the upper right magenta cross.

5. From the View menu (on the Standard AutoCAD menu bar), choose Named Views. The View Control dialog box is displayed.

6. Choose the New button, as shown in the following figure:

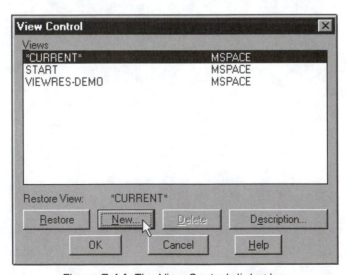

Figure 7-14: The View Control dialog box

7.   The Define New View dialog box is displayed. Enter **stair-11** in the New Name field. The completed Define New View dialog box is displayed, as shown in the following figure. Choose the Save View button.

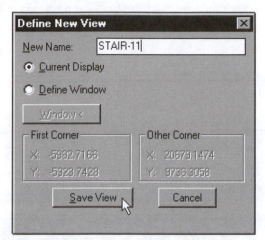

Figure 7-15: The Define New View dialog box

8.   The View Control dialog box is displayed again. Choose **OK.** A view named STAIR-11 is now saved.

9.   In the Aerial View dialog box, select the Pan button. Drag the rectangle over the stairway located at the midway point of the 45 degree exterior wall. The following figure shows the correctly positioned Aerial View rectangle:

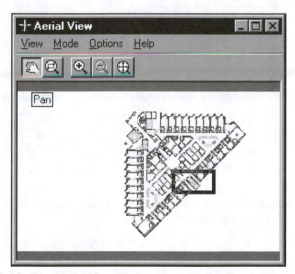

Figure 7-16: The Aerial View dialog box with the Pan button selected

10.   Close the Aerial View dialog box by selecting the X in the upper right corner of the Aerial View window.

You now save the current display as a view. At the Command prompt, enter **view** or its command alias **-v**. Enter **s** for Save and enter the view name, s**tair-12**.

```
Command: -V

VIEW ?/Delete/Restore/Save/Window: S

View name to save: stair-12
```

Working with Viewports

1.   From the View Menu, choose Tiled Viewports, then choose Layout. The Tiled Viewport Layout dialog box is displayed.

2.   Select Four: Equal from the list box. The Tiled Viewport Layout dialog box with Select Four: Equal is shown in the following figure. Choose OK.

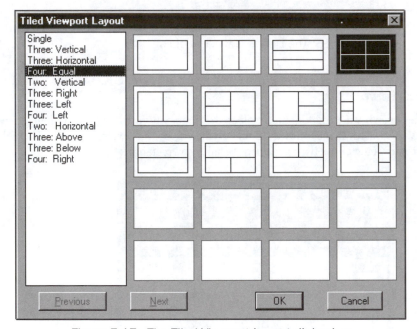

Figure 7-17: The Tiled Viewport Layout dialog box

3.   Four Tiled Viewports are now displayed in the drawing window. Select a point anywhere inside the upper left viewport.

4.   Notice that the upper left viewport border is highlighted, indicating that it is the current viewport.

Now you use the View Control dialog box to restore previously saved views in each viewport.

5. From the View Menu, choose Named Views. In the View Control dialog box, in the Views list box, select STAIR-11. Choose the Restore button. Choose OK. In the current viewport, the view Stair-11 is now displayed.

6. After making each of the following viewports current, recall the DDVIEW command, repeat the command using the following table to restore the corresponding view, and exit the dialog box.

Viewport Description	View Name Description
Upper Right	Stair-12
Lower Left	Viewres-demo
Lower Right	Start

Table 7-1: Description of viewports and corresponding View names

7. Save the drawing as *display1.dwg*. The completed drawing is displayed, as shown in the following figure:

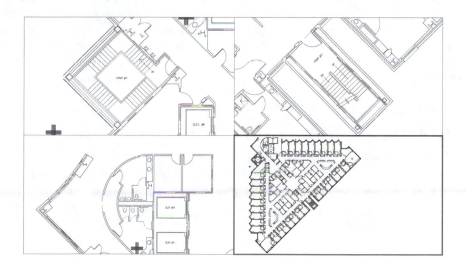

Figure 7-18: The completed display drawing

Refreshing the Drawing Window

The drawing window is updated automatically when certain commands are entered, (for example ZOOM, PAN and VIEW), but there may be times when you intentionally need to refresh the drawing window. The following sections discuss ways to update the drawing window.

Using REDRAW and REDRAWALL

The REDRAW command quickly refreshes the display of the current viewport by removing all blips and any marks left from the use of editing commands. *Blips* are small crosses left on the drawing window after a point is selected. REDRAW is also a transparent command.

Methods for invoking the REDRAW command include:

> ▶ **Menu:** View > Redraw

> ▶ **Command:** REDRAW

The REDRAWALL command functions the same way as the REDRAW command except that it refreshes the graphic areas of all active viewports.
Methods for invoking the REDRAWALL command include:

> ▶ **Toolbar:** Standard

> ▶ **Command:** REDRAWALL

Using REGEN and REGENALL

The REGEN command refreshes the drawing window and recalculates all of the objects in the drawing. When using the REGEN command, it will take AutoCAD a longer time to redisplay your drawing than if the REDRAW command was used. This is a major consideration when working with large drawings. The REGEN command re-indexes the drawing database, re-computes drawing window coordinates for all objects, and smoothes out all circles, arcs, ellipses, and splines. You may need to use REGEN after changes in text styles, layer and linetype properties, and other changes to properties.

Methods for invoking the REGEN command include:

> ▶ **Menu:** View > Regen

> ▶ **Command:** REGEN

The REGEN command only affects the current viewport. If you are working with multiple viewports that need to be updated, use the REGENALL command. The REGENALL command works the same way as REGEN except it regenerates all viewports, and recalculates the drawing window coordinates and view resolution for all objects in each of the active viewports.

Methods for invoking he REGENALL command include:

> ▶ **Menu:** View > Regen All

> ▶ **Command:** REGENALL

Setting View Resolution

The VIEWRES command sets the view resolution. The view resolution refers to the number of lines used to draw circles and arcs. High resolution values display smooth circles and arcs low resolution values display them as short line segments. The only way to invoke the VIEWRES command is from the Command prompt:

When you enter **viewres** at the Command prompt, AutoCAD prompts you as follows:

Do you want fast zooms? <Y>

If you enter **N,** ZOOM, PAN always require a REGEN, and Realtime PAN and ZOOM commands are disabled. If you enter **Y**, AutoCAD performs ZOOM, PAN, and VIEWRES display commands at redraw speed whenever possible. You are then prompted as follows:
Enter circle zoom percent (1-20000) <100>

The circle zoom percent controls the actual smoothness of circles and arcs. The default value is 100 which produces a relatively smooth circle. This value can be set anywhere between 1 and 20000. When numbers larger than 100 are used, the circles and arcs are drawn with fewer vectors (straight lines). If a circle is created using values less than 100 the circle might be displayed as a polygon. When they are drawn with numbers higher than 100, more vectors are used resulting in a smooth circle. The VIEWRES command only affects the way circles and arcs are displayed in the drawing window.   When these objects are printed or plotted they are displayed as smooth continuous arcs or circles.

Exercise 7-3: Updating the Display of Viewports

 In this exercise, you use the REDRAW, REDRAW ALL, REGEN, VIEWPORTS and VIEWRES commands to draw in differing viewports, update the drawing window, and set the view resolution of a drawing.

Drawing Between Viewports

1.   Open the file *regen.dwg*. The drawing has three tiled viewports and looks like the following figure:

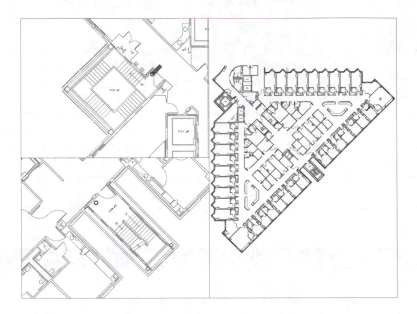

Figure 7-19: Regen.dwg

2. In certain circumstances it is sometimes more efficient to work with multiple viewports. In this exercise, you draw a polyline that starts in one viewport and ends in another viewport.

Position the cursor in the upper left viewport. Press the left mouse button to make the viewport current. The border is highlighted.

3. You now show the rough path a fire protection water pipe takes between stairwells.

Enter **pl** at the Command prompt to start the PLINE command. In response to the PLINE option, From point, select a point near the end of the thick blue polyline that is in the corridor outside of Stair-11.

4. Now make the lower left viewport current.

5. In response to the PLINE option, Endpoint of line, select a point near the center of the blue circle inside Stair - 12. Right-click to exit the command.

Observe that in the right viewport, the Polyline you just made runs from Stair-11 to Stair-12.

Updating the Display

1. In the last part of the exercise when you drew the polyline, blips were left in the drawing window. Now, you remove them using the REDRAW command.

With the lower left viewport current, enter **redraw** at the Command prompt. Observe that the blip in the lower left viewport is gone while the upper left viewport still display its blips.

2. To eliminate blips in all viewports use the REDRAWALL command.

   With the lower left viewport current, enter **redrawall** at the Command prompt. Observe that the blips in the upper left viewport are now gone.

3. Make the right viewport current and use the DDVIEW command to restore the view, VIEWRES-DEMO.

4. Observe that the arcs making up the walls are displayed with line segments. While the right viewport is current, select Redraw from the View menu. Notice the effect that REDRAW had on the arcs that make up the wall.

5. Now you use the REGEN command to update the display of the arcs making up the wall.

   At the Command prompt, enter **regen**. Observe that the arcs that make up the walls are now displayed as smooth arcs.

Setting VIEWRES

In this sub-exercise you see the difference the command VIEWRES makes to arc and circle graphic displays and the related tradeoff in regeneration performance.

1. Make the right viewport current.

2. Enter **vports** at the Command prompt. Then enter **si** for the SINGLE VIEWPORT option.

   Command: **vports**

   Save/Restore/Delete/Join/Single/?/2/<3>/4: **si**

3. At the Command prompt, enter the command **viewres**, and press ENTER. AutoCAD prompts the following:

   Do you want fast zooms?

4. In response to the prompt, enter **n**. AutoCAD prompts the following:

   Enter circle zoom percent (1-20000)

5. In response to the prompt, enter **10**. Observe the display of the arcs (which are line segments) that make up the walls and door swings. Notice the short time required to redisplay the drawing.

6. At the Command prompt, enter the system variable **viewres**, and press ENTER.

7. Enter **yes** at the Command prompt.

8. Enter **20000** at the Command prompt. Observe the smooth display of the arcs that makeup the walls and door swings. Also notice the longer time required to redisplay the drawing.

   Note: Regardless of what value the command VIEWRES is set to, arcs, circles and other curve objects will always plot as smooth curves.

9.   This exercise concludes here. Do not save your drawing.

Conclusion

After completing this chapter, you have learned the following:

▶   When you use drawing or editing commands, you can utilize the transparent PAN and ZOOM commands to change the view and magnification of a drawing.

▶   The Realtime PAN and ZOOM commands let you quickly obtain the display you want.

▶   The Aerial View saves drawing time by displaying your entire drawing in a separate window.

▶   Named Views can be created by using the options in the View Control dialog box.

▶   If you are in the middle of a command and want to use the to change your current view use the ZOOM, PAN, Aerial View, VIEW, and REDRAW commands transparently.

▶   You can create tiled viewports with the VPORTS command to create multiple views of a drawing.

▶   To update or refresh the drawing window, you can use the REDRAW and REGEN commands.

▶   The VIEWRES command sets the view resolution.

Chapter 8

Controlling Object Color and Linetype

Color and linetype are two properties that can be used to differentiate between object sets. This chapter shows you how to change the color of an object and how to edit and modify linetypes.

About This Chapter

In this chapter, you will do the following:

- ▶ Load linetypes and apply linetype scaling.

- ▶ Change object properties with the Object Properties toolbar.

- ▶ Match object properties of different objects.

Color Control - Object Properties Toolbar

To change the color of an object, select the object and then choose a different color from the Color Control drop-down list, as shown in the following figure:

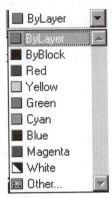

Figure 8-1: The Color Control drop-down list

When the list is closed, the current color selection is displayed. The default listing consists of ByLayer, ByBlock, the seven standard colors, and Other. Selecting Other displays the Select Color dialog box. New colors that you select are added to the bottom of the drop-down list, up to a maximum of four.

Note: Colors are numbered 9-256. Colors 1-8 are named.

During a design session, the compressed Color Control drop-down list displays ByLayer, ByBlock, or the current color. If an object is selected with the system variable PICKFIRST set to 1, then the color of that object is displayed. Selecting multiple objects with different colors displays a blank in Color Control.

Controlling Color Using ByLayer

When you create a new object, AutoCAD assigns the current color. If the current color is set to ByLayer, objects are displayed in the color assigned to the current layer. The default setting for color is ByLayer. If you choose ByLayer, new objects assume the color of the layer upon which they are drawn.

Loading Linetypes

With the Linetype tab of the Layer & Linetype Properties dialog box, you can load linetype definitions from a linetype library file (*.lin), make a linetype current, or globally set linetype scales.

Methods for opening the Layer & Linetype Properties dialog box include:

▶ **Toolbar:** Object Properties

◗ **Menu:** Format > Linetype

◗ **Command:** LINETYPE

The Layer & Linetype Properties dialog box is displayed, as shown in the following figure:

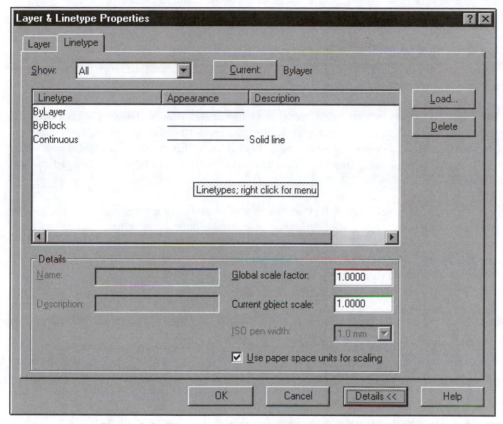

Figure 8-2: The Layer & Linetype Properties dialog box

The Show, Current, and Delete options work the same way as on the Layer tab. You cannot delete linetypes in external references, or linetypes referenced by undeleted layers or objects.

You can load and update the linetypes by choosing the Load button which opens the Load or Reload Linetypes dialog box. You can rename linetypes, except for those that are ByLayer, ByBlock, or Continuous. You cannot rename linetypes in external references. You cannot edit the appearance of the linetype, but you can edit its description in this dialog box. Choose the Linetype column heading to view the loaded linetypes in order.

Choose the Details button to reveal the following three options, which are equivalent to setting these system variables:

Options	Equivalent System Variables
Global scale factor	LTSCALE
Current object scale	CELTSCALE
Use paper space units for Scaling	PSLTSCALE

Table 8-1: Options in the Details section of the Linetype tab

Linetype Drop-Down List in the Object Properties Toolbar

In the Linetype drop-down list on the Object Properties toolbar, you can set the linetype of selected objects. This method is available when Noun/Verb section checkbox is checked in the Object Selection Setting dialog box. The ByLayer and ByBlock options are displayed at the top of the list followed by any other loaded linetypes in alphabetic order. To load other linetypes, choose Load on the Linetype tab of the Layer & Linetype Properties dialog box, or enter **linetype** at the Command prompt. Setting a linetype other than BYLAYER might mean that the selected objects no longer match the layer linetype. This can cause problems in drawing management and is not recommended.

Use LTSCALE to globally set the relative length of dash-dot linetypes per drawing unit. Use PSLTSCALE to control Linetype Scaling in paper space and floating viewports. When set to 0, model space and paper space linetypes are scaled by the global LTSCALE factor. When set to 1 and TILEMODE is set to 0, viewport scaling governs linetype scaling. Dashed length linetypes display at the same scale, in paper space and scaled floating viewports.

Use CELTSCALE to change the linetype scale of selected objects. This setting is the value, times the LTSCALE. An example would be a LTSCALE of 30, selecting an object and setting its CELTSCALE to 2. This yields an object with a Linetype Scale of 60.

Conclusion

After completing this chapter, you have learned:

◗ To change the color of an object, you select the object and then choose a different color from the Color Control drop-down list.

◗ Using the Layer tab of the Layer & Linetype Properties dialog box, you can: change a layer's status, make a layer current, add new layers to the layer name list, rename an existing layer, assign color and linetype to the layers, and delete layers from the list.

Chapter 9

Construction and Adjustable Width Lines

Construction lines are generally used for layouts, projections, and establishing reference points. These lines are created by using the XLINE and RAY commands, and can be manipulated like other objects created with draw commands.   In this chapter, you will learn how to use the XLINE, RAY and PLINE commands.

About This Chapter

In this chapter, you will do the following:

▶ Create construction lines using XLINE and RAY commands.

▶ Draw polylines using the PLINE command.

Creating Construction Lines

Both the XLINE and RAY commands can be used in a similar manner and for similar purposes. The difference is that the XLINE command has more options and is more adaptable than the RAY command. The RAY is also infinite in only one direction.

Construction lines are drawn, plotted, or printed just like any other object. This means you may have problems distinguishing these lines from other lines in the drawing window. To avoid confusion, place construction lines on a separate layer with a different color and distinctive layer name for easy identification.

Using the XLINE Command

The XLINE command creates infinite construction lines. These lines can be edited like other objects created with draw commands. If editing commands are applied to the xline, the type of line will change, for example, if one end of the xline is trimmed, it changes the xline into a ray. This means that it now is a semi-infinite object. If the xline is trimmed in two locations, a line segment is formed.

Methods for invoking the XLINE command include:

- **Toolbar:** Draw

- **Menu:** Draw > Construction Line

- **Command:** XLINE

When you enter **xline**, the following list of options are displayed at the Command prompt. `Hor/Ver/Ang/Bisect/Offset/<From Point>:`. These options are discussed in the following sections.

Two Point

The default two-point option lets you specify two points through which the construction line will pass. When this option is selected, you are prompted to enter a point. This point is referred to as the *root point*. The root point is considered a midpoint and determines the location of the xline on the drawing plane. The second point you select will determine the angle or direction of the xline.

Hor

The horizontal option automatically draws horizontal construction lines through a specified point.

Ver

The vertical option draws vertical construction lines through a specified point.

Ang

The angle option creates construction lines drawn at an angle. There are two ways to create construction lines with the angle option. After you select the angle option, by default you are prompted to enter an angle. After the angular value is entered, you are then prompted to select the point where you want the line to be drawn. The second method lets you create a referenced angled xline. After you select the angle option, select the reference option. This option specifies the angle from the selected reference line, then creates an xline through a specified point using the specified angle.

Bisect

This option creates construction lines that bisect an angle.

Offset

The offset option draws parallel construction lines at a specified distance or through a point. This option operates in the same manner as the OFFSET command discussed in Chapter 9.

Using the RAY Command

The RAY command generally works like the XLINE default except that it extends in one direction. Using the RAY command helps to reduce the number of objects displayed in the display window. After you enter the command, you are prompted to enter an origin point. The second point you select will determine the angle or direction of the ray.

Methods for invoking the RAY command include:

▶   Menu: Draw > Ray

▶   Command: RAY

Exercise 9-1: Using Xlines and Rays as Construction Objects

In this exercise, you use the XLINE and RAY commands to create construction objects. The three viewports shown in the following figure, display the same area, but at different Zoom-Scale factors. First you will use the XLINE command with various options to create lines that extend infinitely in two directions. Then you will use the RAY command to create an object that has a starting point and extends infinitely in one direction. Next you will use the ZOOM command to observe the effects that xlines and rays have on determining drawing extents.

Using the XLINE Command

1.   From the File menu, choose Open, then choose the *ref-line.dwg*. The drawing opens displaying three differently scaled viewports. Use the right viewport for all object selections. The drawing looks like the following figure:

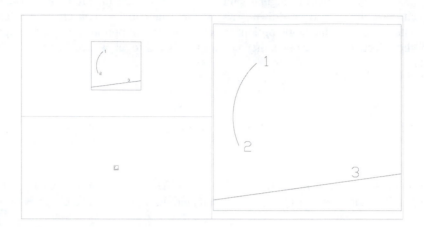

Figure 9-1: The ref-line.dwg

2. From the Draw menu, choose Construction Line.

3. The following XLINE command options are displayed:

   Hor/Ver/Ang/Bisect/Offset/<From point>:

   Enter **v** at the Command prompt to select the Ver (Vertical) command option. For the Through point: enter **600000,0**, then press ENTER two times.

4. Enter the **xline** at the Command prompt. Then enter **h** to select the HOR (Horizontal) command option. For the Through point: enter **0, 600000**, then press ENTER two times.

5. Enter the **xline** at the Command prompt. Then enter **a** to select the ANG (Angle) command option. In response to the Command prompt, Reference/<Enter angle (0)>: enter **45**. For the Through point: enter **600000, 600000**, then press ENTER two times.

6. Enter the **xline** at the Command prompt. Then enter **b** to select the Bisect command option. For the Angle vertex point : use the Intersection object snap, enter **int** and press ENTER, select the intersection of the red xlines

   For the Angle start point : use the object snap Endpoint, enter **end** and press ENTER, select the end of the yellow arc near the number 1. For the Angle end point : use the Endpoint object snap, enter **end** and press ENTER, select the end of the yellow arc near the number 2. Press ENTER to end the command.

7. Press the ENTER to repeat the XLINE command Then enter **o** to select Offset option. The Offset distance or Through point options are displayed. Use the offset distance option, enter **100000**.

   When prompted to Select a line object :, select the yellow line next to the number 3. For the side to offset, select the side the number 3 is on, then press ENTER to end the command.

Using the RAY Command

1. Set the current layer to RAY-OBJECTS. From the Draw menu, choose Ray.

2. In response to the Command prompt, From point : enter **600000,600000**, then press ENTER.

3. For the Through point: use the Endpoint object snap, enter **end** and press ENTER, select the end of the yellow arc near the number 2, then press ENTER to end the RAY command.

Viewing Xlines and Rays

1. Position the cursor in the upper left viewport and press the left mouse button to make the viewport current.

2. From the View menu, Tiled Viewports, then choose 1 Viewport. The tiled viewports have now been replaced by a single viewport.

3. From the View menu, choose Zoom, then choose Extents. When calculating the drawing extents the green rectangle is included, the xlines and rays are not.

4. Now use the ZOOM command alias to zoom out on the drawing. Enter **z**, at the Command prompt. For the Zoom factor, enter **.00000001**. Observe that the green rectangle is barely visible while the xlines and rays extend to the border of the drawing window.

5. The completed drawing looks like the following figure. Do not save the drawing.

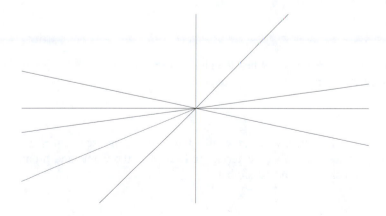

Figure 9-2: The completed ref-line dwg.

Using Polylines

A polyline is a series of connected lines and arcs that are treated as a single object. The PLINE command is used to create polylines. When you use other draw commands, you may find that their capabilities are limited. Polylines, however, are powerful and flexible objects that can represent many drafting situations. With 2D polylines, you can:

- ◗ Create wide and tapered lines

- ◗ Draw contiguous arcs and lines

- ◗ Create polylines of varying widths

- ◗ Create filled circles or donuts

- ◗ Draw closed polygons

- ◗ Modify polylines to create smoothed curves or lines with equally spaced line elements

- ◗ Determine the area and perimeter of closed polyline objects.

Using the PLINE Command

The PLINE command operates in the same manner as the LINE command except that it has more options. All polyline line segments are considered to be one object, can contain curves, and have different widths. The following sections describe some of the polyline options.

Methods for invoking the PLINE command include:

- ◗ **Toolbar:** Draw

- ◗ **Menu:** Draw > Polyline

- ◗ **Command:** PLINE

Close

The Close option closes the polygon. When the close option is selected, a line segment is drawn from the current segment endpoint to the starting point.

Undo

The Undo option lets you remove the most recently drawn arc or line segment. You can remove one line or a series of lines. Do this by entering **undo** or **u** at the Command prompt. You can also use the Undo button on the Standard toolbar.

Width

At times you may find it necessary to change the width of a polyline. To do this, enter **w** at the Command prompt. When the Width option is selected you are prompted to enter a Starting width. After you enter a value for the Starting width you are prompted to enter a Ending width value.

The value you enter for the Starting width becomes the default value for the Ending width. To keep the polyline the same width, except the Ending width default value. Enter different values for the starting and ending widths, if you want the width of the polyline to vary.

Halfwidth

The Halfwidth option lets you draw a polyline using a specified width. The width is taken from the center of a wide polyline and extends to the endpoint of one of its edges. Enter **h** at the Command prompt to access the Halfwidth option.

Arc Options

Arcs are created using the ARC command or using the Arc option from the PLINE command, however, PLINE arcs are more diverse than arcs drawn with the ARC command. When the polyline Arc option is selected, it changes the PLINE command to Arc mode. You are then prompted to choose from a list of sub-options. A description of these sub-options follows:

- *Angle* - Lets you enter the type of angle to be used to specify the span of the arc.

- *Center* - Arcs are generally drawn tangent to other polylines or arcs by default. To establish a location, AutoCAD automatically calculates the center point of each item. The Center option lets you choose a specific location for the center point of the arc. To access this option, enter **ce** at the Command prompt.

- *Close* - Closes the polyline as an arc segment instead of a straight line.

- *Direction* - Lets you choose a specific starting direction for an arc.

- *Halfwidth* - Operates the same way when drawing arcs or line segments.

- *Line* - Switches the PLINE command back to straight line mode.

- *Radius* - Establishes a radius to be used in creating the arc.

- *Second point* - Lets you enter second and third points to draw three points arcs.

- *Undo* - Works in the same manner as the Undo option in straight line mode.

- *Width* - Works in the same manner as the Width option in straight line mode.

Length

You can use the Length option to draw a line segment at a specified length that extends from the last drawn line segment. The new line will be located in the same direction and at the same angle as the previously drawn line segment. Enter **l** to access the Length option.

Exercise 9-2: Drawing Polylines and Arcs

In this exercise, you use the PLINE command to draw an island that divides a parking lot, then draw a direction arrow to illustrate traffic flow. You use PLINE command options along with various forms of coordinate entry to create these objects.

Drawing Polyline Lines and Arcs

1. Open the *parking.dwg* drawing. Verify that Ortho mode is turned on by pressing the **F8** key and observing the <Ortho on> status at the Command prompt. The parking drawing looks like the following figure:

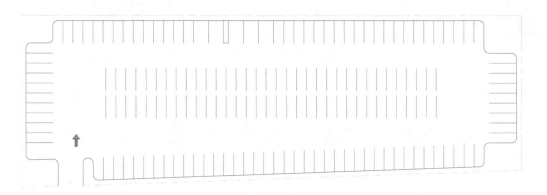

Figure 9-3: parking.dwg

2. From the Draw menu, choose Polyline.

3. For the starting point, enter an absolute coordinate of **624, 936** and press ENTER.

4. The POLYLINE command options are displayed at the Command prompt. For the next point, you will use the direct distance coordinate entry method. Accept the default Polyline <Endpoint of line> command option. Then position the cursor at the bottom of the screen and enter **384**, and press ENTER.

5. Next use the Polyline command option Arc, enter **a** and press ENTER. In response to the prompt <Endpoint of arc>, position the cursor on the right side of the screen and enter **96**, and press ENTER.

6. Next you use the Polyline command option Line, enter **l** and press ENTER. In response to the prompt <Endpoint of Line>, position the cursor at the top of the screen and enter **168**, and press ENTER.

7. For the next point, accept the default Polyline command option <Endpoint of Line>. Position the cursor on the right side of the screen and enter **3360**, and press

ENTER.

8. For the next point, use the relative coordinate entry method. Accept the default Polyline command option <Endpoint of Line>, and enter **@0,-168** and press ENTER.

9. Next use the Polyline command option Arc. Enter **a** and press ENTER. In response to the prompt <Endpoint of arc>, position the cursor on the right side of the screen and enter **96**, and press ENTER.

10. Now use the Polyline command option Line, enter **l**, and press ENTER. In response to the prompt <Endpoint of Line>, position the cursor at the top of the screen and enter **384**, and press ENTER.

11. Next you use the Polyline command option Arc. Enter **a** and press ENTER. In response to the prompt <Endpoint of arc>, position the cursor on the left side of the screen and enter **96**, and press ENTER.

12. Now use the Polyline command option Line, enter **l** and press ENTER. In response to the prompt <Endpoint of Line>, position the cursor directly below the last point entered and enter **168**, press ENTER.

13. For the next point, you use the relative polar coordinate entry method. Accept the default Polyline command option <Endpoint of Line>, and type **@3360<180**, and press ENTER.

14. For the next point, you use the relative polar coordinate entry method. Accept the default Polyline command option <Endpoint of Line>, and type **@168<90**, and press ENTER.

15. For the last point, use the Polyline command option Arc, enter **a**, and press ENTER. In response to the prompt, use the Close option, enter **cl**, and press ENTER.

Drawing POLYLINE with Varying Widths

1. Make the current layer STRIPPING.

    Start the PLINE command with its command alias, enter **pl**, and press ENTER.

2. For the starting point, enter an absolute coordinate of **403,348**, and press ENTER.

3. Next use the Polyline command option Width, enter **w** and press ENTER. In response to the prompt Starting width <0.0000>:, enter **24** and press ENTER.

4. In response to the prompt Ending width <24.0000>:, press ENTER to accept the value inside the brackets.

5. For the next point, use the direct distance coordinate entry method. Accept the current Polyline command option of <Endpoint of line>, position the cursor at the bottom of the screen, and enter **72**, and press ENTER.

6. Next use the Polyline command option Width, enter **w,** and press ENTER. In response to the prompt Starting width <24.0000>:, enter **72,** and press ENTER.

7.   In response to the prompt `Ending width <72.0000>`:, enter **0** and press ENTER.

8.   For the next point, use the direct distance coordinate entry method. Accept the current Polyline command option of `<Endpoint of line>`, position the cursor at the bottom of the screen directly below the last point drawn and type **48**, and press ENTER. Press ENTER again to end the polyline command.

9.   To display the properties of the last object drawn use the LIST command alias, enter **LI**, press ENTER. In response to the prompt, Select Object use the Last option, enter **L** and press ENTER twice . The information displayed indicates the last object drawn was a polyline composed of multiple width segments. Minimize the AutoCAD text window by pressing F2.

10.   The completed drawing should resemble the following figure. Save the drawing *parking.dwg* .

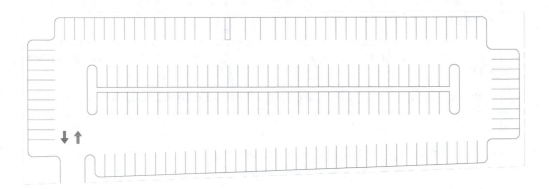

Figure 9-4: The completed parking.dwg

Conclusion

After completing this chapter, you have learned the following:

▶ The XLINE and RAY commands are used to create construction lines. Construction lines are used for layouts, projects and establishing reference points.

▶ The PLINE command operates in the same manner as the LINE command except that it has more command options. All polyline objects created with the PLINE command are considered to be one object.

Chapter 10

Geometric Construction

In this chapter, you learn how to use AutoCAD Draw commands to create objects that can be edited or modified.

About This Chapter

In this chapter, you will do the following:

▶ Use the polygon, rectangle and donut commands to create objects.

▶ Draw objects using the ellipse command.

▶ Draw objects using the spline command.

▶ Use the point command, pdmode, and pdsize system variables to construct points and change point style and size.

Creating Polygon, Rectangle, and Donut Objects

The following sections describe the POLYGON, RECTANGLE, and DONUT commands, and how they are used to create objects.

Using the POLYGON Command

A polygon is a figure with three or more equal sides. AutoCAD lets you create polygons with sides ranging from 3-1024. To create a polygon, use the POLYGON command. The POLYGON command has several options that are used to create objects. The following sections describe these options.

Methods for invoking the POLYGON command include:

> ▶ **Toolbar:** Draw

> ▶ **Menu:** Draw > Polygon

> ▶ **Command:** POLYGON

Center

When you start the POLYGON command, you are prompted to enter its number of sides. You are then prompted to enter the edge or center of the polygon. Center is the default option which lets you draw and size a polygon from its center point. Next, decide if you want your polygon to be inscribed (inside), or circumscribed (outside) of an imaginary circle. After you determine if the polygon will be inscribed or circumscribed, you are prompted to enter a radius.

Inscribed

A polygon is considered inscribed if it is located inside a circle and its corners touch the circle. The following figure illustrates an inscribed polygon:

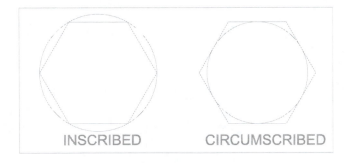

INSCRIBED   CIRCUMSCRIBED

Figure 10-1: Inscribed and Circumscribed polygon options

Circumscribed

The Circumscribed option locates a polygon on the outside of a circle with the sides tangent to the circle. The previous figure illustrates a circumscribed polygon.

Edge

Polygons are created using a center point, by default, however, they can be drawn using an existing edge. To choose this option, enter **e** at the Command prompt.

Using the RECTANGLE Command

Rectangles are created using the RECTANG command. To draw a rectangle select a starting point, then select the opposite corner. You have the option of entering coordinates, or selecting points to achieve the desired size.

Methods for invoking the RECTANG command include:

- ▶ **Toolbar:** Draw

- ▶ **Menu:** Draw > Rectangle

- ▶ **Command:** RECTANG or RECTANGLE

The RECTANGLE command options include:

- ▶ *Chamfer* - Sets the chamfer distance

- ▶ *Elevation* - Specifies the elevation

- ▶ *Fillet* - Specifies the fillet radius

- ▶ *Thickness* - Specifies the thickness of the rectangle.

- ▶ *Width* - Specifies the polyline width

Using the DONUT Command

Donuts are circular polylines created with the DONUT command. You can use this command to draw closed filled circles or rings. You also have the option of entering different values for the inside and outside diameters of the donuts. The system's variable FILLMODE specifies whether multilines, traces, solids, solid-fill hatches, and wide polylines are filled in. If the system variable is set to 0, objects are not filled in. If FILLMODE is set to 1, objects are filled in. The FILL command controls the FILLMODE system variable, and determines visibility of the fill pattern. If FILL is turned off, the donut will be displayed as a series of segmented circles. If FILL is turned on the donut fill pattern will be visible. The following figure illustrates donuts with FILL turn off and on.

Methods for invoking the DONUT command include:

- ▶ **Menu:** Draw > Donut

- ▶ **Command:** DONUT

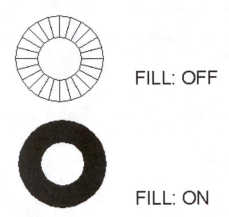

FILL: OFF

FILL: ON

Figure 10-2: Donuts with the FILL command on and off options selected

Creating Ellipses

AutoCAD can be used to draw ellipses and elliptical arcs. An ellipse is a circle that can be viewed from various angles. All ellipses have a center, major axis, and minor axis. The major axis of an ellipse is longer than the minor axis. The following sections describe how the ELLIPSE command and options operate.

Using the ELLIPSE Command

The ELLIPSE command uses the PELLIPSE system variable, which determines the type of ellipse that will be created. The default value for the PELLIPSE is 0. This creates true elliptical objects. To draw a polyline representation of an ellipse, enter 1 for the value.

Methods for invoking the ELLIPSE command include:

- **Toolbar:** Draw
- **Menu:** Draw > Ellipse
- **Command:** ELLIPSE

When you access the ELLIPSE command, you are given options for the method of creating your ellipse. These are the Arc, Center, and Axis Endpoint options, covered in the following sections.

Arc

Elliptical arcs are drawn by selecting the Arc option. They are generally drawn in the same manner as ellipses created with the Center and Axis Endpoints options. However, the Arc option has a few additional sub-options. A description of these options follows:

The following sequence creates an arc using default settings: When you select the Arc option **a,** you are prompted to:

1. Enter Axis endpoint 1

2. Axis endpoint 2

3. Choose either Other axis distance, or Rotation.

4. Select the Other axis distance default option.

5. After you enter the values at the command prompt for these options, an ellipse will be displayed. The center point of the ellipse, and the degree value you enter, or point you choose for the start angle, will be used to determine the starting point of the arc.

6. Then you are prompted to enter an end angle which will determine the length of the arc. This is done by entering a degree at the prompt or dragging the cursor until you achieve the desired length, then press ENTER.

The Parameter option operates in the same manner as the default setting options, except it uses a different system to calculate vectors. This option also uses a different command sequence. Instead of entering values for the start and end angles, enter the same information at the start parameter and end parameter prompts. To access this option, enter **p** when the Parameter option is displayed at the Command prompt.

The Included Angle option locates an included angle at the beginning of the start angle. The Include Angle option operates like the previously discussed options except that you enter **i** at the Parameter/Included<end angle> Command prompt.

The Rotate option lets you rotate the elliptical arc around the first axis by entering a rotation angle. The rotation option is used in the same manner as the rotate option for full ellipses. The Center option command option lets you create an elliptical arc using an established center point. To access this option enter **c** at the <Axis endpoint 1>Center Command prompt.

Center

The center point of an ellipse is located where the major and minor axes cross. The Center option lets you draw an ellipse by specifying the center point and the endpoints of two axes. To draw an ellipse with the center option, enter **c** at the Command prompt. Then locate a point in the Drawing window to represent the center point of the ellipse. You are then prompted to locate the Axis endpoint. The Axis endpoint is selected by choosing a point in the drawing window or entering a number. You are then prompted to choose the default Other axis distance or Rotate to complete the command sequence. If you choose the default option you can enter a number at the Command prompt, or select a distance by choosing a point. You can also choose the Rotate option. You can manually rotate the ellipse around the major axis or enter a rotation angle between 0-89.4 degrees. An angle larger than 89.4 degrees will display the ellipse as a line.

Axes Endpoints

This default option lets you create an ellipse by establishing the major and minor axes. When this option is selected, you are prompted to locate the first axis endpoint, then select the second endpoint. The major and minor axes are defined by the values you enter at the next prompt. You then have the option of choosing the Other axis distance default, or Rotation. Both of these sub-options work the same way for the Center and axis endpoint options.

Exercise 10-1: Drawing Polygons, Rectangles, Donuts, and Ellipses

In this exercise, you use the POLYGON, RECTANGLE, DONUT and ELLIPSE commands to draw whirlpool fixtures.

Creating a Rectangle

1. From the File menu, choose New.

2. Choose the Start From Scratch button, under the Select Default Settings list, select Metric, then choose OK.

3. From the Draw menu, choose Rectangle.

4. Enter **0,0** at the `First corner:` Command prompt. Then press ENTER.

5. Enter **96,60** at the `Other corner:` Command prompt. Then press ENTER twice to repeat the command.

6. Enter **2,29** at the `First corner:` Command prompt

7. Enter **10,31** at the `Other corner:` Command prompt. Then press ENTER.

8. From the View menu, choose Zoom, then choose Extents.

Using the Ellipse command

1. To draw an ellipse, enter **el** at the Command prompt.

2. Enter **5,30** at the `Arc/Center/<Axis endpoint 1>:` Command prompt. Then press ENTER

3. Enter **91,30** at the `Axis endpoint 2:` Command prompt, then press enter.

4. Enter **25** at the `<Other axis distance>/Rotation:` Command prompt. Then press ENTER. The drawing should resemble the following figure:

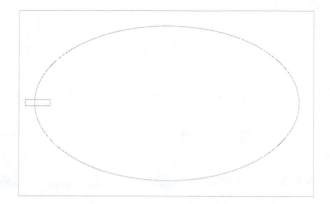

Figure 10-3: The object after using the RECTANGLE and ELLIPSE command

Using the POLYGON command

1. From the Draw menu, choose POLYGON.

2. At the Number of sides <4>: Command prompt, enter **6,** then enter, **3,24** at the Edge/<Center of polygon>: Command prompt.

3. Enter **i** for inscribed, then enter **2** for radius. Press ENTER twice.

4. At the Number of sides <4>: Command prompt enter **6,** then enter, **3,36** at the Edge/<Center of polygon>: Command prompt.

5 Enter **i** for inscribed, then enter **2** for radius. Then press ENTER.

Using the DONUT command

1. From the Draw menu, choose DONUT.

2. At the Inside diameter: Command prompt, enter, **2.5**.

3. Enter **3** at the Outside diameter: Command prompt.

4. At the Center of doughnut: Command prompt, enter **48,30**. Then press ENTER. The completed drawing looks like the following figure:

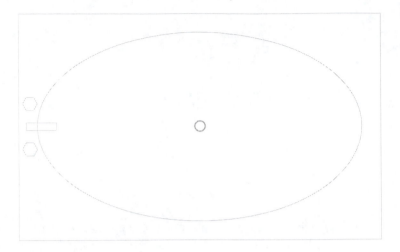

Figure 10-4: Completed exercise

Creating Splines

A spline is a curved line that passes through a series of specified points. Edited polylines can be used to create linear approximations of splines, it is better to use the SPLINE command to draw a true spline. The SPLINE command uses NURBS (Non-Uniform Rational B-Spline) to create true curved lines or objects. Coordinates are used to locate the points through which the spline will pass.

Curves can be created using a variety of commands that include CIRCLE, ARC, ELLIPSE, DONUT, PLINE and SPLINE. There are however, certain advantages in using the SPLINE command.

- ▶ Using the SPLINE command will save disk space and the memory of your computer.

- ▶ The SPLINE command lets you change spline fitted polylines into true splines.

- ▶ When SPLFRAME system variable is set to 1, AutoCAD will display spline-fit polylines and polygons and the control points used to create the spline objects. Setting the SPLFRAME to 0 turns off the frame.

- ▶ Displayed polylines and polygons are straight line representations of spline objects.

Using the SPLINE Command

The SPLINE command is accessed by selecting SPLINE from the draw menu, entering SPLINE at the Command prompt, or by selecting the SPLINE icon. When you start the command, you are given the choice to draw objects using the spline default settings. The default options let you select points in the drawing window or enter coordinates at the Command prompts. The Start tangent determines the direction of the line created at the starting point. The End tangent determines the direction of the spline endpoint.

Methods for invoking the SPLINE command include:

> ▶ **Toolbar:** Draw

> ▶ **Menu:** Draw > Spline

> ▶ **Command:** SPLINE

Object

The Object option lets you change a 2-D spline fitted polyline into a true spline. To change the spline fitted polyline into a true spline, choose the SPLINE command Object option, then select the spline fitted polyline.

Close

The Close option closes the spline curve by defining the last point as coincident with the first and making it tangent to the joint. This option is selected by entering **c** at the `Close/Fit Tolerance/<Enter Point>:` Command prompt. After you select the Close option, a line is drawn from the endpoint to the starting point of the spline. You are then prompted to `Enter tangent:`. This option lets you manipulate the direction of the line used to close the spline object.

Fit Tolerance

A spline can be drawn with a set of fixed points. Instead of using fixed points, you may want to draw a spline within a set boundary or region. When the tolerance is set to 0, the spline will pass directly through specified points. If the tolerance value is set higher than 0, the spline will be drawn in the region between the points. Higher numbers will generate the spline further away from the specified points. The Fit Tolerance option is invoked by entering **f** at the `Close/Fit Tolerance/<Enter Point>:` Command prompt.

Exercise 10-2: Creating and Editing Splines

In this exercise, you select a series of points to create a pond. You use the SPLINE command, SPLFRAME system variable, and grips to draw and modify this object.

Creating a Spline

1. Open the file *pond.dwg,* then choose Spline from the Draw menu.

2. Select the staring point by placing the cursor over the blue point on the left side of the Drawing window, then press the left mouse button.

3. To create the spline, use the following sequence to select the remaining points. Position the cursor, then choose the red, green, yellow, cyan, white, and magenta points.

4. To close the spline, enter **c** at the Command prompt.

5. To choose the tangent point, then select the magenta point. The completed spline is shown in following figure:

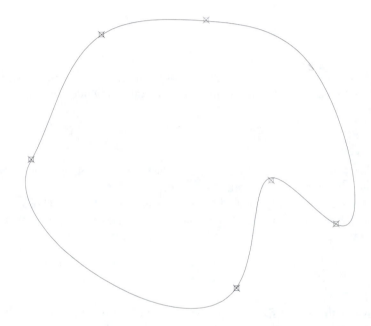

Figure 10-5: Completed Spline

Editing a Spline

1. Use the Layer Control drop down list to freeze the POINTS layer.

2. Place the cursor on the boundary of the spline, then press the left mouse button. Note all of the spline grips (small squares) that control the shape of the spline are now visible.

> Note: To turn on the Grips feature, set the System Variable Grips to 1. At the command prompt enter Grips, and then enter a value of 1.

3. Position the cursor over the grip on the left side of the drawing window, then press the left mouse button. Note how the grip changed from an outline to a solid red square.

4. To reposition the spline, use the cursor to select a location to the far left of the drawing window, then press the enter button. The spline outline is now updated based on the new grip location.

5. To remove the grips from the display, press ESC twice. The updated spline should be

similar to the following figure:

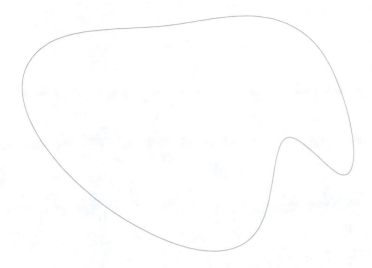

Figure 10-6: The modified spline

6. You will now use the system variable SPLFRAME to view the spline control polygon from which the pond was created. At the Command prompt, enter **splframe**, then enter a value of **1**.

7. Select REGEN from the View menu, or enter the keyboard alias **re** at the Command prompt. The spline control polygon is displayed in the following figure. Excercise 5 is now complete. Do not save your drawing.

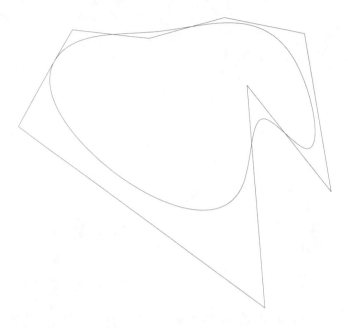

Figure 10-7: Spline showing pline control polygon

Placing Points

Many AutoCAD drawing and editing commands will require you to specify the location of points in the drawing window. You can use the left mouse button to place points in the drawing window, or use one of the coordinate entry methods. You can also place a point by using object snaps.

Using the POINT Command

Some drawing and editing commands let you place points in the drawing window by responding to the Command prompts located within the commands. The POINT command also lets you place points in the drawing window. The POINT command options include single point which places one point on the screen, multiple point which draws more than one point consecutively. The MEASURE and DIVIDE commands can place points along lines and arcs.

Methods for invoking the POINT command include:

> ▶ **Toolbar:** Draw

> ▶ **Menu:** Draw > Point

> ▶ **Command:** POINT

Point Style Dialog Box

The PDMODE and PDSIZE system variables are used to modify the appearance of points. Use the PDMODE system variable, or the Point Style dialog box to change the point style. Each point style is assigned a number. The PDMODE lets you change the style by entering a number at the Command prompt. To see the various point options, select Point Style from the Format menu. The Point Style dialog box is displayed which lets you choose the style you want. When you use the PDMODE to select a new point style, you must perform a regeneration. Use the REGEN command to see the changes.

The point size is set by the PDSIZE system variable, or by entering information in the Point Size area of the Point Style dialog box. When you select the Set Size Relative to Screen from the Point Size area, the point size will change to a percentage relative to the display window. This means if you zoom in or zoom out the point size will change along with the other objects that are modified. The Set Size in absolute Units will keep the point size set at the same size. For example, if you zoom in or out the point will stay at the set size. This means your drawing could be displayed with extremely large points or points that are too small to see. A regeneration must also be performed after changing the point size.

Methods for opening the Point Style dialog box include:

▶ **Menu:** Format > Point Style

▶ **Command:** DDTYPE

The Point Style dialog box is displayed in the following figure:

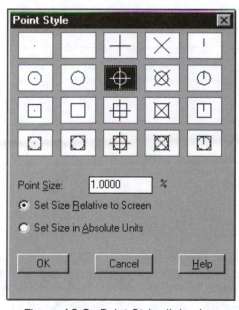

Figure 10-8: Point Style dialog box

Using the DIVIDE and MEASURE Commands

The DIVIDE command lets you divide a linear object into equal parts. However, you can not use the DIVIDE command with blocks, dimensions, text, or hatch. When you enter the DIVIDE command, you are prompted to Select object to divide:. Then enter the number of segments at the `<Number of segment>/Block:` Command prompt. The markers are then placed on the selected object. The appearance of the markers is set by the Point Style settings or the block you selected.

Methods for invoking the DIVIDE command include:

- ▶ **Menu:** Draw > Point > Divide

- ▶ **Command:** DIVIDE

The MEASURE command measures the length in specified units along a linear object. These unit values are set at the Command prompt. When you use the MEASURE command, points will be placed along the object that represent one unit. If AutoCAD reaches an area on the object that is not a full unit, it will not place a point. The appearance of the markers is set by the Point Style settings or the block you selected.

Methods for invoking the MEASURE command include:

- ▶ **Menu:** Draw > Point > Measure

- ▶ **Command:** MEASURE

The following figure illustrates points placed using the DIVIDE and MEASURE commands:

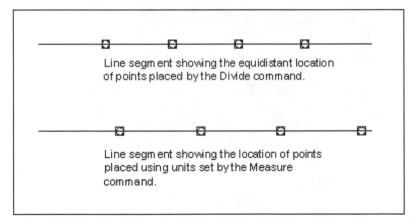

Line segment showing the equidistant location of points placed by the Divide command.

Line segment showing the location of points placed using units set by the Measure command.

Figure 10-9: Two Lines modified with the Divide and Measure commands.

Exercise 10-3: Drawing Points and Changing the Point Style and Size

A series of speakers need to be placed in the corridor of a building. In this exercise you use the POINT command, modify the point style, use REGEN to update your drawing, and use the DIVIDE and MEASURE commands to identify the speaker locations.

Modifying Points

1.   Open the file *speaker.dwg*, the drawing looks like the following figure:

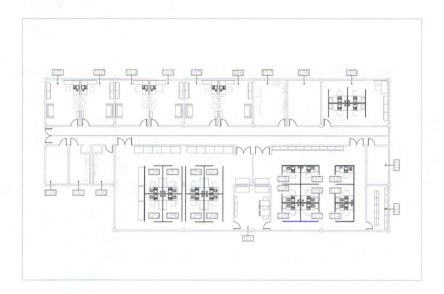

Figure 10-10: Speaker.dwg

2.   Enter the POINT command alias **po** at the Command prompt. In response to the Point prompt, enter the absolute coordinate **450,400**, then press ENTER. Notice the appearance of the red point in the bottom left of the drawing.

3.   From the Draw menu, choose Point, then choose Divide. In response to the Select object to divide: prompt, choose the yellow corridor line. Then enter **15** for the Number of segments. The drawing looks like the following figure:

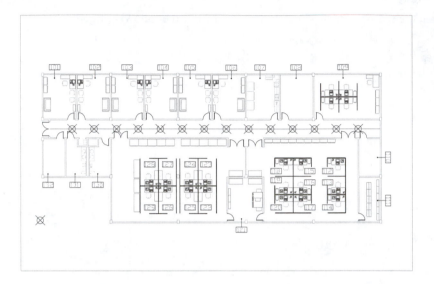

Figure 10-11: Speaker.dwg showing points on a divided line

4.   Use the Layer Control drop down list to make the MEASURE layer current.

5.   From the Format menu, choose Point Style. The Point Style dialog box is displayed.
        Select the point style located in the third row, fourth column. Then choose OK. The
        Point Style dialog box is shown in the following figure:

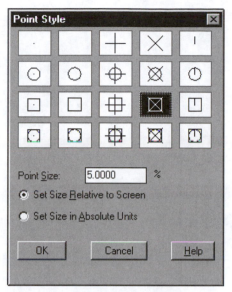

Figure 10-12: Point Style dialog box

6. From the Draw menu, choose Point, then choose Measure. Select the yellow corridor line, type **400** for the `Segment length` , then press ENTER. Notice the points are now displayed with the new point style.

7. From the View menu, choose Regen to update the drawing window. Notice all of the points now look the same.

8. Enter **view** at the Command prompt to display the View Control dialog box. Choose CORE from the Views area, select Restore, then choose OK. Enter the REGEN command, to update the drawing window. Notice the size of the points which are displayed at 5%, relative to the drawing window.

9. From the Format menu, select Point Style. The Point Style dialog box is displayed. Set the Point Size to **40**, choose the Set Size in Absolute Units button, then choose OK. Enter **regen** at the Command prompt. Note all displayed points have a size of 40 units.

10. From the View menu, select Zoom, then Previous. Then enter **regen** at the Command prompt. Note all displayed points still have a size of 40 units. This completes exercise 6, do not save your drawing. The completed drawing looks like the following figure:

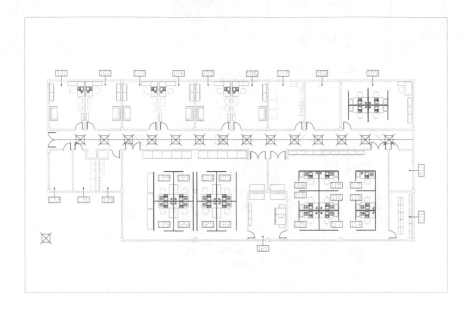

Figure 10-13: The completed Speaker.dwg

Conclusion

After completing this chapter, you have learned the following:

▶ The POLYGON command lets you draw objects with 3 or more equal sides.

▶ The RECTANGLE command is used to draw rectangles by entering two points which determine the length and height of the rectangle.

▶ The DONUT command is used to draw objects.

▶ The ELLIPSE command and PELLIPSE system variable are used to determine the type of ellipse or elliptical arcs that will be drawn.

▶ The SPLINE command is used to draw curved lines that pass through a series of specified points.

▶ To create points in the drawing window, use the POINT command, and use the Point Style dialog box to control the appearance of points.

▶ You have learned how to use the DIVIDE and MEASURE commands.

Chapter 11

Locating Geometric Points

In this chapter, you learn how Object Snaps serve an important function helping to ensure quality and accuracy in the drawing process. You also learn how Object Snaps let you reference specific locations on selected objects quickly and easily.

About This Chapter

In this chapter, you will do the following:

- ▶ Observe the importance of using Object Snaps.

- ▶ Practice procedures for using Object Snaps.

- ▶ Define Running Object Snaps.

- ▶ Override Running Object Snaps.

- ▶ Practice controlling the AutoSnap feature.

Using Object Snaps - Geometric Points

Object Snaps are used to locate an exact position on an object without the need to know the absolute coordinate or draw construction lines. Using object snaps is faster than referencing drawing point locations using coordinate values.

Select an Object Snap mode whenever AutoCAD prompts for a point. You can access object snaps from the Object Snap toolbar, from the Object Snap flyout on the Standard toolbar, or from the Cursor menu (SHIFT + right-click). Selecting a single button from the Object Snap toolbar affects only the next object you select.

The Object Snap toolbar is shown in the following figure:

Figure 11-1: Object Snap toolbar

Using Endpoint and Midpoint Osnaps

Endpoint snaps to the closest endpoint of an arc, elliptical arc, line, multiline, polyline segment, or ray. You select the object at a location closest to the end to which you want to snap. Endpoint also snaps to the closest corner of a trace, solid or 3Dface.

Methods for invoking the Endpoint object snap include:

> ▶ **Toolbar:** Object Snap, Standard

> ▶ **Menu:** Cursor

> ▶ **Command:** END

Midpoint snaps to the midpoint of an arc, elliptical arc, line, multiline, polyline segment, solid, spline, or xline. Objects can be selected at any location.

Methods for invoking the Midpoint object snap include:

> ▶ **Toolbar:** Object Snap, Standard

> ▶ **Menu:** Cursor

> ▶ **Command:** MID

The following figure shows the AutoCAD cursor placed on a line or an arc to reference the endpoints or midpoints of the objects.

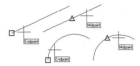

Figure 11-2: Endpoint and Midpoint with AutoSnap Markers and Snaptip

Note: If the AutoSnap™ system variable is enabled, your cursor automatically locks onto the snap location you select. The AutoSnap™ feature instantly displays the appropriate Marker. If you hold the cursor over the object, a Snaptip is displayed which informs you of the object snap that is being applied. If AutoSnap™ is disabled, you will not receive a visual cue that you are snapping to the point you have selected.

Using Center and Quadrant Osnaps

Center snaps to the center of an arc, circle, ellipse, or elliptical arc. To select the object, select the line that defines the object. For example, if you are trying to snap to the center of a circle, you must select the actual line that defines the circle and not the blank area at the center of the circle.

Methods for invoking the Center object snap include:

> **Toolbar:** Object Snap, Standard

> **Menu:** Cursor

> **Command:** CEN

Quadrant snaps to a quadrant point of an arc, circle, ellipse, or elliptical arc. The quadrant points on an object are the absolute 0, 90, 180, and 270 degree locations defined by the current user coordinate system (UCS).

Methods for invoking the Quadrant object snap include:

> **Toolbar:** Object Snap, Standard

> **Menu:** Cursor

> **Command:** QUA

The following figure shows how the AutoCAD cursor must be placed on the line of the object in order for the object snap to be applied and for the AutoSnap Marker and Snaptip to be displayed.

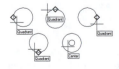

Figure 11-3: Center and Quadrant object snaps

Using Node and Insertion

Node snaps to a point object.

Methods for invoking the Node object snap include:

▶ Toolbar: Object Snap, Standard

▶ Menu: Cursor

▶ Command: NODE

> Note: The Point style may have to be changed to make it easier to identify point locations when the points are drawn on an object by the DIVIDE or MEASURE commands.

Insertion snaps to the insertion point used to place an attribute, a block, an image, a shape, or text. Methods for invoking the Insertion object snap include:

▶ **Toolbar:** Object Snap, Standard

▶ **Menu:** Cursor

▶ **Command:** INS

The Node AutoSnap marker is displayed at the center of the point. The Insertion marker is the "square-above-square" graphic above the text in the following figure:

Figure 11-4: Node and Insertion object snaps

Using Intersection and Apparent Intersection Osnaps

Intersection snaps to the intersection of an arc, circle, ellipse, elliptical arc, line, multiline, polyline, ray, spline, or xline, and another object of the same or different type.

Methods for invoking the Intersection object snap include:

▶ **Toolbar:** Object Snap, Standard

▶ **Menu:** Cursor

▶ **Command:** INT

You may also snap to the Extended Intersection of two objects with the Intersection object snap. AutoCAD lets you use the Extended Intersection snap to select the imaginary intersection of two objects that would intersect if they were extended along their natural paths. Extended Intersection is automatically enabled when you select the Intersection object snap mode.

Extended Intersection is invoked when the aperture box is placed over only one object. If AutoSnap markers are checked on, AutoCAD will display the Intersection marker, followed by three periods. After the object is selected, you are prompted to select a second object with:

        _int of and

Upon selection of the second object, AutoCAD snaps to the imaginary or apparent intersection formed by extensions of the two selected objects.

> Note: Care must be taken when approaching a group of objects for selection. If there is more than one object in the aperture box when you are prompted to select the second object, AutoCAD will use the first object it finds. This may not be the object you wanted to select.
>
> If the Intersection and Apparent Intersection running object snaps are both enabled, you may get varying results when you attempt to select an intersection.

Apparent Intersection snaps to the apparent intersection of two objects (arc, circle, ellipse, elliptical arc, line, multiline, polyline, ray, spline, or xline), that do not physically intersect in 3D space. Apparent Intersection cannot be used for objects that are parallel because they will not intersect anywhere in 3D space.

Methods for invoking the Apparent Intersection object snap include:

▶ **Toolbar:** Object Snap, Standard

▶ **Menu:** Cursor

▶ **Command:** APPINT or APP

The following figure displays the Intersection snap marker at the intersection of the two diagonal lines on the left. The Extended Apparent Intersection snap point (the X on the lower right) shows the intersection point of two selected objects if they were extended along their natural paths. The Extended Apparent Intersection mode requires two objects to be chosen before the snap point is displayed.

Figure 11-5: Intersection and Apparent Intersection object snaps

Perpendicular and Tangent

Perpendicular snaps to a point perpendicular to an arc, circle, ellipse, elliptical arc, line, multiline, polyline, ray, solid, spline, or xline. You can also use the arc, circle, line, multiline, polyline, ray, xline, or 3D solid edge as objects from which to draw a perpendicular line.

Methods for invoking the Perpendicular object snap include:

> ▶ **Toolbar:** Object Snap, Standard
>
> ▶ **Menu:** Cursor
>
> ▶ **Command:** PER

Deferred Perpendicular snap mode (DPSM) can be used to draw perpendicular lines between objects. DPSM is automatically enabled when you use the edge of a line, arc, circle, pline, ray, xline, mline, or 3D solid as the first snap point from which to draw a perpendicular line. If the AutoSnap perpendicular marker is enabled, the marker will be displayed when you select the first point. You are then prompted to enter a second point. The line is then drawn perpendicular to the object chosen as the first snap point.

> Note: Deferred Perpendicular object snap does not work with ellipses or splines. If you draw a line that is perpendicular to an ellipse or spline, you will get the point on the ellipse or spline that is perpendicular to the last point chosen. This can cause unpredictable results.

Tangent snaps to the tangent of an arc, circle, ellipse, or elliptical arc.

Methods for invoking the Tangent object snap include:

> ▶ **Toolbar:** Object Snap, Standard
>
> ▶ **Menu:** Cursor
>
> ▶ **Command:** TAN

The Deferred Tangent snap mode (DTSM) can be used when more than one tangent is to be completed. DTSM is automatically enabled when you select an arc, pline arc, or circle as a starting point for a tangent line. If AutoSnap is enabled, you will see the deferred tangent Snaptip and marker appear when you select the first point, and you will be prompted to enter a second point. The line is then drawn tangent to the object chosen as the first snap point.

> Note: DTSM do not work with ellipses or splines. If you draw a line that is tangent to an ellipse or spline, you will get the point on the ellipse or spline that is tangent to the last point picked. This can cause unpredictable results.

The bold line in the following figure was created using the Tangent and Perpendicular object snap modes:

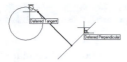

Figure 11- 6: Perpendicular and Tangent object snaps

Nearest

Nearest snaps to the point nearest the center of the cursor on an arc, circle, ellipse, elliptical arc, line, multiline, point, polyline, ray, spline, or xline.

Methods for invoking the Nearest object snap include:

▶ **Toolbar:** Object Snap, Standard

▶ **Menu:** Cursor

▶ **Command:** NEA

Note: The only accuracy that the Nearest object snap has is that it will select a point on the object selected. The point will only be nearest the location of the cursor at the time of selection.

The bold line in the following figure was drawn from the point nearest the selection of the circle and the point nearest the selection of the line:

Figure 11-7: The Nearest object snap

Quick

Quick snaps to the first point on the first object found. Quick must be used in conjunction with other object snap modes. Quick is especially useful when you are snapping to an intersection where there are several objects intersecting at that same location.

Methods for invoking the Quick object snap include:

▶ **Toolbar:** Object Snap, Standard

▶ **Menu:** Cursor

▶ **Command:** QUI

Exercise 11-1: Using Object Snaps

Object snaps are an essential part of creating an accurate drawing. This exercise focuses on the use of the most common object snaps. In this exercise, you access the Object Snap modes from the Object Snap toolbar.

Using Object Snaps

1. Open the file *osnap_01.dwg*. The drawing looks like the following figure:

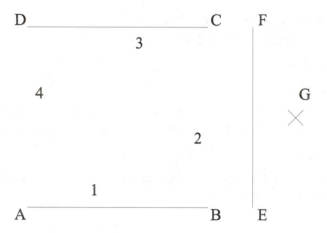

Figure 11-8: Osnap_01.dwg

2. From the View menu, choose Toolbars. In the Toolbars dialog box, in the Toolbars drop-down list, check the Object Snap checkbox to activate the Object Snap toolbar (*see Figure 8-1*). Close the Toolbars dialog box.

Using Endpoint and Intersection

1. From the Draw toolbar choose the Line button. From the Object Snap toolbar, select the Snap to Endpoint button. Then select the endpoint of line AB near A.

2. Select the endpoints of the following lines in sequence choosing the Snap to Endpoint button before each selection:

    ▶ Line CD near C

    ▶ Line AB near B

    ▶ Line CD near D

Then enter **c** at the last *to point* Command prompt.

3.   From the Draw toolbar, choose the Circle button. Then choose the Snap to Intersection button from the Object Snap toolbar. Select the intersection of lines AC and DB.

4.   Enter **1** at the Command prompt for the radius and press ENTER.

Using Quadrant and Perpendicular

1.   From the Draw toolbar, choose the Line button.

2.   From the Object Snap toolbar, choose the Snap to Quadrant button. Then select the circle in area 1. Select Snap to Perpendicular from the Object Snap buttons and select line BD.

      Complete the command by pressing ENTER.

3.   Repeat steps 1 and 2 for the following areas and lines:

      ◗   Quadrant of area 2 and Perpendicular to line AC

      ◗   Quadrant of area 3 and Perpendicular to line BD

      ◗   Quadrant of area 4 and Perpendicular to line AC

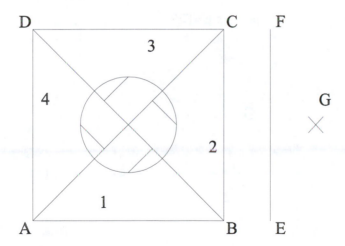

Figure 11-9: The osnap_01.dwg after step 3

Using Center and Midpoint

1.   From the Draw toolbar, choose the Line button.

2.   From the Object Snap toolbar, select the Snap to Center button. Then select the circle.

Select Snap to Midpoint from the Object Snap buttons and select line AB.

Complete the command by pressing ENTER.

3.   Repeat steps 1 and 2 for the following:

▶   Select the circle and line BC

▶   Select the circle and line CD

▶   Select the circle and line AD

Using Node, Endpoint, and Midpoint

1.   From the Draw toolbar, choose the Line button.

2.   From the Object Snap toolbar, select the Snap to Node button. Then select the Point G. Select Snap to Midpoint from the Object Snap buttons and select line EF.

Complete the command by pressing ENTER.

3.   From the Draw toolbar, choose the Line button.

4.   From the Object Snap toolbar, select the Snap to Endpoint button and choose line EF near E. Select the Snap to Node button from the Object Snap toolbar, then select the Point G. Select Snap to Endpoint from the Object Snap buttons and select line EF near F.

Complete the command by pressing ENTER.

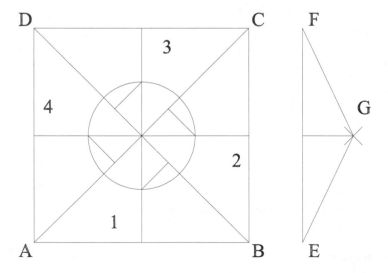

Figure 11- 10: Step 4 completed

Using Extended Intersection, Perpendicular, Tangent, Nearest, and Insert

1. From the Draw toolbar, choose the Line button.

2. From the Object Snap toolbar, select the Snap to Intersection button. Select line AC, next position the cursor over line EF, then move the cursor along the axis of line EF. When the Intersection marker is displayed, press the left mouse button.

3. For the next point, use the Object Snap cursor menu to define your snap pick mode. To do this hold down the SHIFT key and the press the right mouse key. The Object Snap cursor menu is displayed, select Perpendicular and then select line AD. Complete the command by pressing ENTER.

4. From the Draw toolbar, choose the Line button.

5. From the Object Snap toolbar, select the Snap to Tangent button and the circle in area 4. Select the Snap to Nearest button from the Object Snap toolbar, then toward the left end, select the line created in step 2-3.

   Complete the command by pressing ENTER.

6. Press ENTER to repeat the LINE command. From the Object Snap toolbar, select the Snap to Tangent button and the circle in area 2. Select the Snap to Nearest button from the Object Snap toolbar, then toward the right end, select the line created in step 2-3.

   Complete the command by pressing ENTER.

7. Press ENTER to repeat the LINE command.

8. From the Object Snap toolbar, select the Snap to Insert button. Then select the text on A.

9. Select the insertion of the following text in sequence, choosing the Snap to Insert button before each selection:

   ▶ B

   ▶ C

   ▶ D

   To complete this series of line segments, use the command option close. In response to the command prompt, *to point*, enter **c**. You drawing should look similar to the following figure. This concludes the exercise, do not save your drawing.

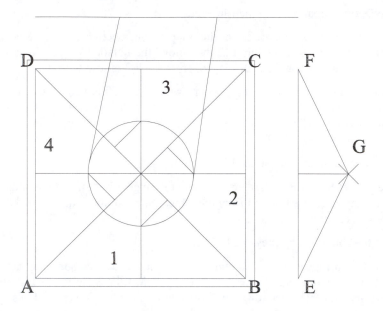

Figure 11- 11: Completed osnap_01.dwg file

Running and Override Object Snaps

During the drawing process, you may want to select a specific object snap for a series of objects. Running object snaps let you set single, as well as multiple object snaps. They are called "running" because they are "locked on" and ready for use whenever an object can be selected.

> Note: An Aperture box is displayed at the center of the cursor to indicate that an object snap is on and to identify the selection area. This applies to both individually selected object snaps and running object snaps.

> The APERTURE system variable sets the size of the Aperture box up to 50 pixels. The Aperture size setting is stored in the system registry, and will stay in effect until you specify another size. You can also change the size of the Aperture box in the Osnap Settings dialog box (see Figure 8-13).

Selection Methods

Object snaps play such an important part in creating a fast and accurate drawing that AutoCAD provides more ways of controlling them than almost any other feature in the software.

Methods for invoking the Osnap Settings dialog box include:

◗ **Toolbar:** Object Snap, Standard

◗ **Menu:** Tools > Object Snap Settings
   Cursor menu (SHIFT + right-click) > Osnap Settings

◗ **Command:** OSNAP

Once you have defined a running object snap, you can quickly toggle object snaps on and off without having to redefine your object snap settings. This is done by double-clicking on Osnap in the Status bar, (see the following figure) pressing CTRL + F, or pressing F3.

SNAP GRID ORTHO OSNAP MODEL TILE

Figure 11- 12: Status Bar

Running Osnap Tab in the Osnap Settings Dialog Box

The Osnap Settings dialog box contains two tabs which are Running Osnap and AutoSnap(TM), as shown in the following figure:

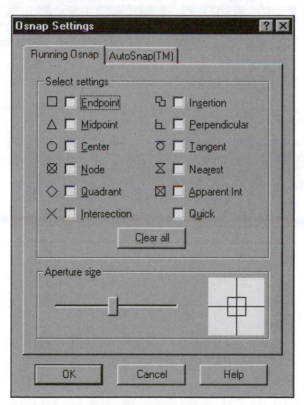

Figure 11-13: Osnap Settings dialog box with Running Osnap tab chosen

In the Select Settings area of the Running Osnap tab, single and multiple running osnaps are set by checking the checkbox next to the appropriate object snap(s). A selected running osnap can be turned off by clearing the individual checkbox. The Clear all button is used to quickly turn off all running osnaps.

When you select multiple object snaps, AutoCAD uses the one object snap that is appropriate to the object you select. If more than one potential snap point falls within the selection area, AutoCAD snaps to the eligible point closest to the center of the Aperture box.
A graphic of the AutoSnap Markers for each of the object snap modes is shown next to each checkbox in the dialog box.

Note: The From object snap cannot be set as a running object snap.

The Aperture size area of the Running Osnap tab is used to specify the size of the aperture box. Drag the slider bar from left to right to determine a desired aperture box size.
The aperture box image tile changes to show the new size. The maximum size is 20 pixels using the Osnap Settings dialog box.

Exercise 11-2: Using Running and Override Object Snaps

In this exercise, you define multiple running object snaps and then apply them to the drawing process.

Using Running Object Snaps

1. Open the file *osnap_02.dwg*. The drawing looks like the following figure. This drawing needs to have several circles connected with lines to define cutting edges for trimming.

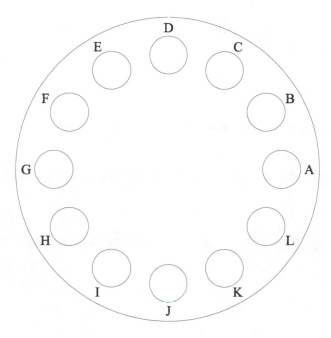

Figure 11-14: Osnap_02.dwg

2. On the Status bar, double-click Osnap to access the Osnap Settings dialog box.

3. Check the Center object snap checkbox and choose OK.

4. Enter **line** at the Command prompt and begin drawing a line from the center of circle A. Complete the LINE command by connecting to the centers of each circle from A to B, B to C, continuing on to L. After you draw the line from the center of K to L, enter **c** for Close. Press ENTER.

   By setting the Center running object snap, it is much quicker to complete the drawing task than if you have to select the Center object snap button individually for each circle.

5. This completes the running object snap exercise. Your drawing should looks like the following figure, do not save your drawing.

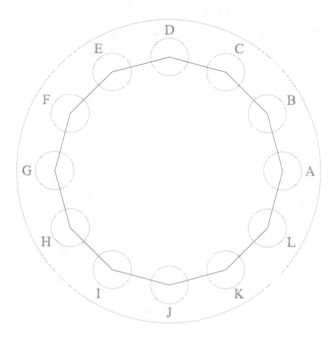

Figure 11-15: The completed Osnap_02.dwg

Setting Running Object Snaps

1.   Open the file *osnap_03.dwg*. The drawing looks like the following figure. Here you draw additional lines using Endpoint and Midpoint running object snaps.

Figure 11-16: The osnap_03.dwg file

2. Select the Object Snap Settings button from the Object Snap flyout on the Standard toolbar. The Osnap Settings dialog box is displayed.

3. Check on the Endpoint, Midpoint, and Perpendicular checkboxes and choose OK.

4. Choose the LINE command and begin a line from the Midpoint of line AB. Continue the line vertically across to the Midpoint of line CD. Press ENTER.

> Note: The Perpendicular marker may not be displayed unless you cycle through the available object snap modes by using the TAB.

5. Press ENTER to repeat the LINE command and draw another line horizontally across the midpoints of lines BC and AD. Press ENTER.

6. Choose the Line button from the Draw toolbar.

7. In this exercise step it should be noted that a point entered with absolute coordinates has priority over active running object snaps.

    Enter **1,0** at the From point: Command prompt and press ENTER. Place the cursor on line CD and then move the cursor along the line until the Perpendicular marker is displayed. Then select the line.

    Complete the command by pressing ENTER.

Using Object Snap Overrides

1. Use the CIRCLE command to draw a circle at the coordinate of 7,2 with a radius of 1.

2. Choose the Line button from the Draw toolbar.

3.   In response to the `From point:` Command prompt, select the endpoint of the lines located next to letter B.

4.   Now position the cursor on the edge of the circle closest to letter B. Observe the Perpendicular object snap tooltip, this is the only valid running object snap. Select the circle.

5.   In response to the `To point:` Command prompt, enter **cen** and press ENTER. In response to the prompt `Cen of`, position the cursor over the edge of the circle, and select the circle.

By entering the required snap mode at the Command prompt, the current running object snap settings were overridden.

6.   In response to the `To point:` Command prompt, press and hold the SHIFT key, then press the right mouse button. The Object Snap cursor menu is displayed.

7.   From the Object Snap cursor menu, select Insert. Position the cursor over the letter C. When the tooltip Insert marker is displayed, select letter C. Press the right mouse button to end the LINE command.

When the Insert object snap was selected from the Object Snap cursor menu, the current running object snap settings were overridden.

Changing Priority for Coordinate Data Entry

1.   You now change the Priority for Coordinate Data Entry from Keyboard Entry Except Scripts to Running Object Snap.

From the Tool menu, select Preferences. The Preferences dialog box is displayed.

2.   Choose the Compatibility tab. In the Priority for Coordinate Entry Data area, choose the Running Object Snap checkbox. Then choose OK.

With this setting selected, object snaps always have priority when points are entered.

3.   Start the LINE command. Enter **2,0** at the `From point:` Command prompt and press ENTER. Place the cursor in the center of the circle, and look for the Endpoint marker, then press the left mouse button.

Complete the command by pressing the right mouse button.

The line should have snapped to the midpoint of line AB even though the coordinate location for the midpoint of line AB is 3,0. This is because the running object snap settings are now overriding the coordinate input.

4.   Choose the Line button from the Draw toolbar.

5.   Double click the Osnap button on the Status bar. This toggles the running object snaps off. Enter **2,0** at the `From point:` Command prompt and press ENTER.

6.   Double-click on the Osnap button on the Status bar. This toggles the running object snaps back on. Place the cursor on line CD and look for the Perpendicular marker, then

select the line.

Complete the command by pressing the right mouse button.

7.   Now change the system variable that controls Priority for Coordinate Data Entry back to
      its original default setting. At the Command prompt, enter **osnapcoord**. In response to
      the prompt `New value for OSNAPCOORD <0>:`, enter **2**.

8.   This completes the overriding running object snap exercise. Your drawing should look
      like the following figure, do not save your drawing.

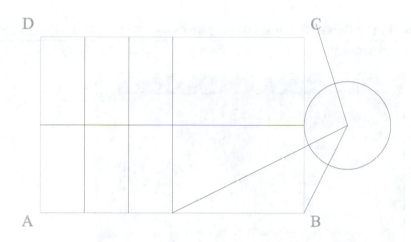

Figure 11-17: The completed Osnap_03.dwg

Note: The OSNAPCOORD system variable controls whether coordinates
entered at the Command prompt override running object snaps. The
OSNAPCOORD system variable can also be set in the Preferences dialog
box, under the Compatibility tab. The following are accepted values:

▶   0 - Running object snap settings override keyboard coordinate entry

▶   1 - Keyboard entry overrides object snap settings

▶   2 - Keyboard entry overrides object snap settings except in scripts

Using AutoSnap

Default settings automatically turn on AUTOSNAP™ when you enter an object snap is entered at
the Command prompt, or when running object snaps in the Osnap Settings dialog box are turned
on.

When you select any of the object snap settings and drag the cursor over an object, AutoSnap™
displays a marker. If you hold your cursor over a snap point displaying the marker, a Snaptip is

displayed.

The AUTOSNAP™ system variable controls the display of the AutoSnap marker and Snaptips, and turns the AutoSnap Magnet on or off. Values entered are the sum of the following bit values:

- ◗ *0* - Turns off the Marker, Snaptip, and Magnet

- ◗ *1* - Turns on the Marker

- ◗ *2* - Turns on the Snaptip

- ◗ *4* - Turns on the Magnet

AutoSnap Tab in the Osnap Settings Dialog Box

The following figure shows the AutoSnap(TM) tab on the Osnap Settings dialog box:

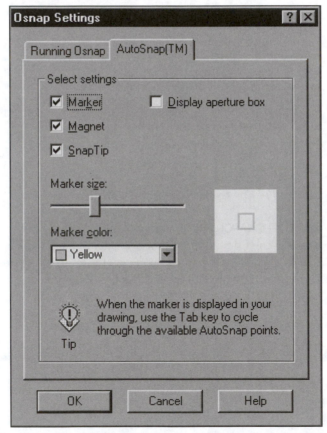

Figure 11-18: AutoSnap tab in the Osnap Settings dialog box

The available Select settings are described as follows:

- ◗ *Marker* - This checkbox turns the Marker on or off. The Marker is a geometric shape that is used to indicate the type of object snap being applied and the location on the object. Markers are displayed when your aperture box moves over an object.

▶ *Magnet* - This checkbox turns the Magnet on or off. The Magnet locks your aperture box onto the snap point much like snapping to a grid point.

▶ *Snaptip* - This checkbox Turns the Snaptip on or off. The Snaptip is a text description of the object snap mode being applied to the snap location.

▶ *Display Aperture Box* - Controls the display of the aperture box at the center of the cursor when object snap modes are selected. This checkbox turns the aperture box on or off.

▶ *Marker Size* - Controls the size of the Marker in pixels. Drag the slider bar left or right to decrease or increase the size of the Marker.

▶ *Marker Color* - Shows the current color of the Marker. You can change the color by selecting a new color from the Marker Color drop-down list.

▶ *Image Tile* - Shows the current size and color of the Marker.

Note: As the tip explains, you can use the TAB key to cycle through the available AutoSnap™ points in the dialog box. You can also cycle through all the snap points available for a particular object by pressing the TAB key as you hold your cursor on an object. Objects are highlighted during the cycling process.

Conclusion

After completing this chapter, you have learned the following:

▶ Object snaps are used to help ensure quality and accuracy in the drawing process.

▶ Object Snap modes are accessed from the Object Snap toolbar, from the Object Snap flyout on the Standard toolbar, or from the Cursor menu (SHIFT + right-click).

▶ The size of the aperture box may be adjusted using the Osnap Settings dialog box, or the APERTURE system variable.

▶ Define running object snaps for repetitive drawing task.

▶ Running Object Snap modes can be toggled on and off by using the OSNAP button from the Status menu, F3 and CTRL+F.

▶ The AutoSnap features *Marker*, *Snaptips*, and *Magnet* may be set with the AutoSnap tab in the Osnap Settings dialog box.

Chapter 12

Modify Commands - Location and Form of Objects

In this chapter, you learn how modifying in AutoCAD is easier than conventional drawing, and re-drawing repetitive symbols and line work is no longer necessary. You learn how to use the AutoCAD Modify commands to simplify the drawing process, increase drawing productivity, logically structure your drawing work, speed up the flow of drawings, and avoid repetitive drawing work.

About This Chapter

In this chapter, you will do the following:

◗ Create and use selection sets.

◗ Modify object locations and properties with the modify commands.

Using Modify Commands

Modify commands offer shortcuts in the drawing process. Increased speed and efficiency encourages designers to spend more time on accuracy, and to concentrate on the design and engineering aspects of the project.

You can carry out a wide variety of actions with the Modify commands. When these commands create new objects, they create them on the same layer and with the same properties as the "parent" object. Current layer, linetype, and color settings are ignored.

Creating and Using Selection Sets

When using editing commands like ERASE, MOVE, COPY and MIRROR, AutoCAD lets you select more than one object at a time. This collection of objects is known as a *selection set*. AutoCAD has several different selection methods available.

At the `Select objects` prompt of a command, you can choose any object to form the selection set by selecting it. At the same time AutoCAD prompts you to select objects, the cursor is replaced by a pickbox. You place the pickbox cursor on an object and press the left mouse button to select the object. The object display changes to a dashed line, known as a highlight, to indicate that it has been selected.

You can also create a selection set by using a technique known as *implied windowing*. At the `Select objects` prompt, select a point in your drawing which is not on an object, then drag the cursor and select another point to form a box. If you drag to the left, the box is displayed as a dashed line, and is termed a *crossing* selection. If you drag to the right, the box is displayed as solid lines, and is termed a *window* selection. A crossing selection includes objects that are within the box and which cross any of the edges of the box, while a window selection only includes objects completely within the box.

Other selection methods include the following (shortened Command prompt entry is shown in brackets):

- *All[all]* - Selects all objects in the drawing, except for those on frozen or locked layers.

- *Window[w]* - Selects objects enclosed in a window. At the prompt, drag a rectangle around the required objects in any direction.

- *Crossing[c]* - Acts exactly as the window option, but includes any object within or passing through the window.

- *Wpolygon[wp]* - Lets you select objects as with the window option, except that the window is a closed (irregular) polygon boundary.

- *Cpolygon[cp]* - Acts exactly as the WPolygon option, but uses the crossing window capability.

- *Fence[f]* - Lets you create a polyline. Any objects that are crossed over are selected.

- ▶ *Last[l]* - Automatically selects the last object drawn
- ▶ *Previous[p]* - Selects objects in the most recent selection set
- ▶ *Single[si]* - Forces the selection of a single object
- ▶ *Multiple[m]* - Lets you select objects without highlighting them. In large drawing files, selecting objects can cause AutoCAD to pause while it locates and highlights each object.

After you use any of the above methods to create a selection set, AutoCAD reports how many objects you have added to the selection. To complete the selection of objects, press ENTER at the Select objects prompt. The following is an example of a typical editing operation involving selecting objects:

Command: **ERASE**

Select objects: 1 found (user selected one object with the pointing device)

Select objects: Other corner: 61 found (user used implied windowing)

Select objects: **f**   (the Fence selection set)

First fence point:

Undo/<Endpoint of line>:

Undo/<Endpoint of line>: (user presses ENTER)

6 found

Select objects: (user presses ENTER to end the selection process and execute the command)

Add and Remove Options

If you include an object in a selection set which you do not want, enter **r** (**remove**), and select the objects to remove, by using any of the selection methods. You can also remove objects by selecting them while pressing the SHIFT key. AutoCAD reports how many objects it found from your remove operation.

To undo the last selection in a selection set, enter **u** (**undo**). If you want to return to adding objects to the selection set, enter **a** (**add**), and continue selecting the objects to be included. The following is an example of using the remove option to take objects out of a selection set:

Command: **erase**

Select objects: Other corner: (user adds to selection set with implied windowing)

29 found

Select objects: **r**

Remove objects: **wp** (user uses a window polygon to select objects to remove from selection set)

First polygon point:

Undo/<Endpoint of line>:

Undo/<Endpoint of line>:

Undo/<Endpoint of line>:   (user presses ENTER to complete this selection)

19 found, 18 removed (19 objects were found, and 18 were in the selection set - these are removed from the selection set)

Remove objects: (user presses ENTER to end the selection process and execute the command)

> Note: If the HIGHLIGHT system variable is set to 0, objects are not highlighted as they are added to a selection set.

Using the ERASE Command

The ERASE command removes selected objects from the drawing, and is probably the easiest editing command to use.

Methods for invoking the ERASE command include:

> ▶ **Toolbar:** Modify

> ▶ **Menu:** Modify > Erase

> ▶ **Command:** ERASE

Once the command is started, choose the objects you want to erase. You can either select objects individually, or use any of the selection set methods previously explained. When you are finished, press ENTER to complete the selection and carry out the command.

Command: **erase**

Select objects: (1 found)

Select objects: (press ENTER)

Using the UNDO, REDO and OOPS Commands

You can easily undo your most recent action or actions. Use the U command to undo a single action and the UNDO command to undo several actions. For example, you can enter a specific number of actions to undo or use the Mark option of the UNDO command to mark an action as you work. You can then undo back to that action with the Back option.

Methods for invoking the UNDO command include:

> ▶ Toolbar: Standard

> ▶ Menu: Edit > Undo

> ▶ Commands: UNDO, U

To undo the most recent action, enter **U** at the Command prompt. This undoes the last action you performed, such as deleting objects, creating a new layer, or defining a dimension.

You cannot undo commands such as SAVE, OPEN, NEW or COPYCLIP that write data to, or read data from, a disk. To undo a multiple series of actions, you can use the UNDO command. For example, to undo the last seven actions, enter **7**. AutoCAD shows the actions that are undone. The undone commands and actions are shown at the Command prompt.

There are several more advanced options in the UNDO command. The UNDO Command prompt is the following:

Auto/Control/BEgin/End/Mark/Back/<Number>:

A description of these options follows:

> ▶ *Auto* - Undoes a menu selection as a single command, reversible by a single U command.

> ▶ *Control* - Limits or turns off the UNDO command. You can specify the number of operations. Specifying a small number reduces the amount of memory or hard disk used by the UNDO command, therefore improving performance.

> ▶ *BEgin and End* - The Begin option groups a sequence of operations. All subsequent operations become part of the group until End terminates the group.

> ▶ *Mark and Back* - The Mark option places a mark in the undo information. The Back option undoes all the work completed up to this mark. If you undo one operation at a time, AutoCAD informs you when you reach the mark.

You can place as many marks as necessary. Back undoes work by moving back one mark at a time, removing the mark. If no mark is found, Back shows this prompt:

This will undo everything. OK? <Y>   (Enter **y** or **n,** or press ENTER)

Entering **y** undoes everything done since starting AutoCAD or working on the current drawing. Entering **n** disables the Back option. A mark stops multiple UNDO operations if the number entered is greater than the number of operations since the mark.

For more details on these options, see the online *AutoCAD Command Reference.*

You can redo any action undone with the U and UNDO commands by using the REDO command. You can only redo an action immediately after undoing it.

Methods for invoking the REDO command include:

> ▶ Toolbar: Standard

> ▶ Menu: Edit > Redo

> ▶ Command: REDO

The OOPS command, accessed only at the Command prompt, is a refined version of the UNDO command. The OOPS command undoes the last command that removed data from your drawing. For example, if you entered this sequence of commands:

- ▶ Erase

- ▶ Line

- ▶ Arc

- ▶ Layer

and then entered **oops**, the command would not only restore data deleted with the ERASE command, but would also retain objects created with the LINE and ARC commands, and the changes from the LAYER command. The command is normally used after you create a block with the BLOCK command. OOPS puts the geometry from the block back into the drawing.

Using the MOVE and COPY Commands

The MOVE command lets you move one or more objects.

Methods for invoking the MOVE command include:

- ▶ **Toolbar:** Modify

- ▶ **Menu:** Modify > Move

- ▶ **Command:** MOVE

The prompts for the MOVE command are as follows:

```
Command: Move
Select objects: (Select object(s) you want to move)
Select objects: (Press ENTER)
Base point or displacement: (Select a point)
Second point of displacement: (Select a point)
```

Before moving the object or objects, AutoCAD prompts for a *base or reference point*. You can enter this as an absolute coordinate or as a location specified by an object snap. You also need to input a *destination point* to inform AutoCAD how the object or objects has to be moved. For example, if you want to move a selection set by 10 units in the X direction, and 10 units in the Y direction, you could enter **0,0** at the first prompt, and **10,10** at the second. You could also choose a location on an object, and show the place to which you want the object moved.

For example, in the following figure, you want to put a corner of the rectangle at the center of the circle:

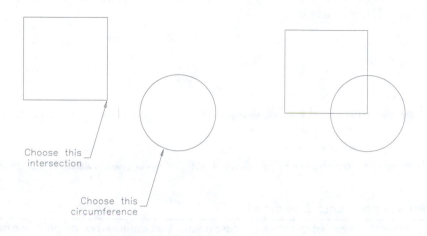

Choose this
intersection

Choose this
circumference

Figure 12-1: Moving a circle using object snap

An example of using the MOVE command prompts is as follows:

Command: **move**

Select objects: 1 found (select the rectangle)

Select objects: (Press ENTER)

Base point or displacement: **int** *of* (Select the corner of the rectangle)

Second point of displacement: **center** *of* (Select any point on the circle circumference)

If you press ENTER at the Second point of displacement prompt, the first entry is used as a relative displacement. In this example, the selection set moves by 10 units in the X direction and 10 units in the Y direction.

Base point or displacement: **10,10**
Second point of displacement: (Press ENTER)

The COPY command works in the same way as the MOVE command, except that the selected objects are both left in their original positions, and are copied to the new location that you specify.

Methods for invoking the COPY command include:

▶ **Toolbar:** Modify

▶ **Menu:** Modify > Copy

▶ **Command:** COPY

If you are creating more than one copy of a selection set, use the Multiple option at the <Base

point or displacement>/Multiple prompt. When you enter **m**, the command prompts you for the displacement again, and then lets you continue making copies until you press ENTER, as in the following example:

```
Command: copy
Select objects: Other corner: 2 found
Select objects:
<Base point or displacement>/Multiple: m
Base point or displacement:
Second point of displacement:
Second point of displacement:
Second point of displacement: (press ENTER to end the command)
```

Using the ROTATE and SCALE Commands

The ROTATE command moves objects around a base point. You define a central point for rotation and then specify the angle. Remember that by default, positive rotation is in a counter-clockwise direction.

Use the ROTATE command to modify existing objects.

Methods for invoking the ROTATE command include:

- ▶ Toolbar: Modify
- ▶ Menu: Modify > Rotate
- ▶ Command: ROTATE

An example of the ROTATE command prompts is as follows:

```
Command: rotate
Select objects:
Base point:
<Rotation angle>/Reference:
```

The Rotation Angle lets you specify an absolute angle to rotate the selected objects. The Reference option lets you indicate an existing angle and then the angle which you want the selection set to use. This is a good way to re-orient objects, or align an object with another.

The following figure shows a sample rotation with the Base point at the lower corner of the diamond shape:

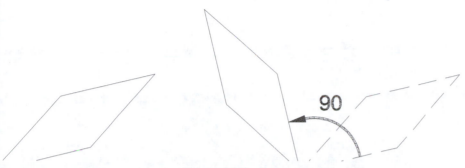

Figure 12-2: Example of a rotation

The SCALE command enlarges or reduces selected objects by equal amounts in the X and Y directions.

Methods for invoking the SCALE command include:

▶ **Toolbar:** Modify

▶ **Menu:** Modify > Scale

▶ **Command:** SCALE

The following list describes the available options of the SCALE command:

▶ *Base Point* - Indicates the origin point for the scaling. You might select a corner of an object or the center of an object, as shown in Figure 9-3.

▶ *Reference Point* - Lets you specify the existing scale or size and define a new value. For example, if an object has a scale of 10 and you want it to have a scale of 5, you enter **10** and **5** at the prompts for Reference Length and New Length.

▶ *Scale Factor* - Increases or decreases the amount of the scale. Scale factors larger than 1 increase the size, while scale sizes smaller than 1 decrease the size.

The following figure displays two examples of using the SCALE command:

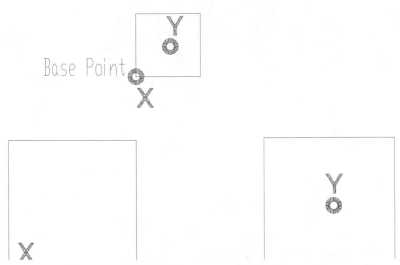

Figure 12-3: Using the SCALE command with different base points

Using the STRETCH Command

The STRETCH command moves or stretches objects.

Methods for invoking the STRETCH command include:

> ▌ **Toolbar:** Modify

> ▌ **Menu:** Modify > Stretch

> ▌ **Command:** STRETCH

AutoCAD stretches lines, arcs, elliptical arcs, splines, rays and polyline segments that cross the selection window. STRETCH moves the endpoints that lie inside the window, leaving those outside the window unchanged. STRETCH also moves vertices of traces and solids that lie inside the window and leaves those outside unchanged. Polylines are handled segment-by-segment, as if they were primitive lines or arcs. STRETCH does not modify polyline width, tangent, or curve-fitting information.

You must use the Crossing (C) or Crossing Polygon (CP) options to select the objects. Any objects entirely within the crossing window or polygon are moved, as if you were using the MOVE command. Objects that cross the selection box area are stretched.

When you specify how objects will be displaced, AutoCAD displays the following prompts:

Base point or displacement: (Specify a point or press ENTER)
Second point of displacement: (Specify a point or press ENTER)

If you enter a second point, the objects are stretched the vector distance from the base point to the second point. If you press ENTER at the Second point of displacement prompt, STRETCH treats the first point as the X,Y displacement value.

The following figure is an example of using the STRETCH command:

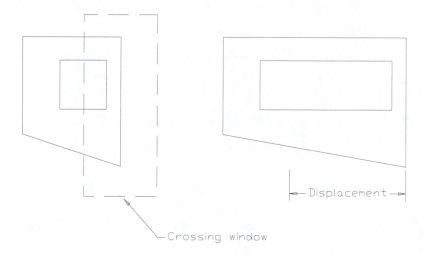

Figure 12-4: Using the STRETCH command

Exercise 12-1: Using the Move, Copy, and Stretch Commands

In this exercise, you use the MOVE, COPY, and STRETCH commands to edit and complete a simple mechanical drawing of a plate with bolt holes. You also define and work with selection sets, and use reference points to specify edit operations precisely. Lastly, you use the ERASE command to remove some geometry from the drawing.

Using MOVE and COPY

1.   Open the file *mechpart.dwg* The drawing looks like the following figure:

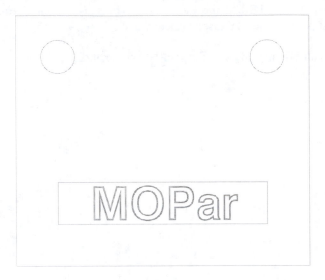

Figure 12-5: mechpart.dwg

2.   Now you will move the nameplate and associated text to the middle of the plate.

      From the Modify menu, choose Move.

3.   At the `Select objects` prompt, use your cursor to create an implied window. To do this, select the point shown by P1 in the following figure, and then move the cursor to P2 and select again:

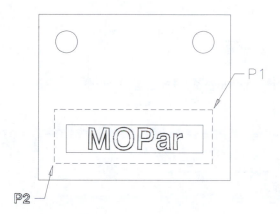

Figure 12-6: Points to select for the MOVE command

4.   Press ENTER to complete the selection set.

Now you need to specify the displacement. In this case, you move the selection with a positive Y displacement of 0.75.

5.   At the `Base point or displacement` prompt, enter **0,0.75**. At the `Second point of displacement` prompt, press ENTER.

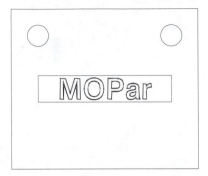

Figure 12-7: Part after MOVE completed

6. Now copy the top bolt holes to the bottom of the plate. From the Modify menu, choose Copy.

7. Select the two bolt holes with an implied window.

8. For the base point, use an intersection object snap to select the top left corner of the plate. Then use direct distance entry with Ortho mode on to enter a displacement of **2** toward the lower edge of the plate.

Your Command prompt should look like this:

Command: **copy**

Select objects: Other corner: 2 found

Select objects:

<Base point or displacement>/Multiple: **int** *of* (choose intersection)

Second point of displacement: *<Ortho on>* **2**

The plate looks like the following figure:

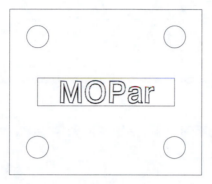

Figure 12-8: Plate after copying bolt holes

Using STRETCH to Lengthen the Plate

1.　Now stretch the plate, including the bolt holes and nameplate text. At the Command prompt, enter **stretch**.

2.　At the Select objects prompt, use an implied crossing window to include the right-hand side of the plate and at least the midpoint of the text, then press ENTER. The drawing looks like the following figure:

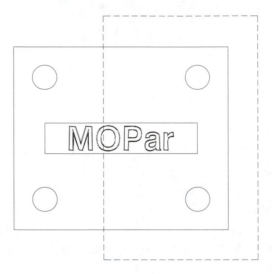

Figure 12-9: Implied window to select objects

3.   Use an intersection osnap to select the bottom left corner of the plate as the Base point, and specify a displacement of **5** units in the +X direction.

The stretched plate looks like the following figure:

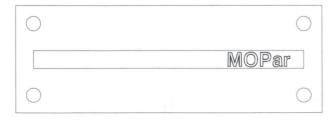

Figure 12-10: Stretched plate

If you did not include the middle of the text in your selection window, the text will remain on the left side of the plate, rather than being moved to the right.

4. Enter **erase** at the Command prompt to remove the text from the drawing. At the `Select objects` prompt, select any point on the text. Then press ENTER.

Do not save the changes to your drawing.

Using the ALIGN Command

The ALIGN command is a useful way to move and rotate objects at the same time. The way that the command works varies according to how many source and destination points you input, and whether you work in two or three dimensions.

Methods of invoking the ALIGN command include:

   ▶ **Menu:** Modify > Align

   ▶ **Command:** ALIGN

The ALIGN Command prompt looks like this:

```
Command: align
Select objects:
1st source point: 1st destination point:
2nd source point: 2nd destination point:
3rd source point: 3rd destination point:
```

If you specify one source and destination point and then press ENTER, the command will work the same way as the MOVE command.

If you specify two source and destination points, as shown in the Figure 9-11, you can either align in 2 or 3 dimensions. In Figure 9-11, the points S1 and S2 are the first and second source points, and D1 and D2 are the destination points respectively. If you choose a 2D transformation, the command performs a move and rotate operation on the selection set. The original object is scaled or stretched if you enter **yes** at this prompt:

```
Scale objects to alignment points? [Yes/No] <No>:
```

The following figure is an example of using the ALIGN command. The second object has been scaled but the third has not:

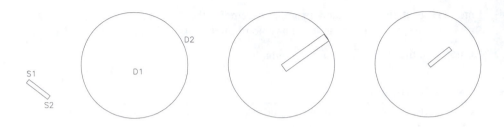

Figure 12-11: Using the ALIGN command

Using the MIRROR Command

The MIRROR command lets you mirror objects around a mirror line. This is useful when creating objects or geometry that is symmetrical about an axis.

- ▶ **Toolbar:** Modify

- ▶ **Menu:** Mirror

- ▶ **Command:** MIRROR

The prompts for the MIRROR command are:

```
Command: Mirror
Select objects: (Pick object(s))
Select objects: (Press Enter)
```

After selecting the object(s) to be mirrored, indicate the first position for the mirror line. The mirror line is the axis about which the object(s) are mirrored over. The next option lets you retain the original object(s) selected for mirroring.

```
First point of mirror line:
Second point of mirror line:
Delete old objects? <N>
```

> Note: If text is selected as an object to be mirrored, then you can set the variable MIRRTEXT to 0 (off) which prevents the text from being mirrored.

Using the OFFSET Command

The OFFSET command creates a new object, not just a copy of an existing object. This new object is not only copied from the original object, but may be adjusted to compensate for the distance drawn away. If the object contains a filleted corner, the fillet radius is adjusted automatically to take into account the distance away from the original object.

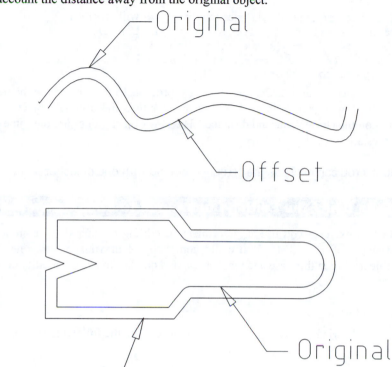

Figure 12-12: Examples of using OFFSET

Methods for invoking the OFFSET command include:

▶ **Toolbar:** Modify > Offset

▶ **Menu:** Modify > Offset

▶ **Command:** OFFSET

The prompts for the OFFSET command are:

Command: **Offset**
Offset distance or Through <Through>:

At this prompt you have two choices:

Use Through to select the object and then indicate the point the copy must pass through.
Select object to offset: (select the object)
Through point: (select a point or enter distance)

Select the object to offset:

Use Offset Distance to set the distance that the object will be offset.
Offset distance: (select a point or enter distance)
Select object to offset: (select the object)
Side to offset ? (select a point on one side of the object)
Select the object to offset:

Use the Through point and Side to offset prompts to indicate where the new object is to be drawn. The new object is drawn larger or smaller than the original object it was copied from, depending on the chosen side and distance. The new object's lines are the same size, but its polylines and arcs are resized.

Note: You can only use the OFFSET command with one object at a time.

Exercise 12-2: Using the Editing Commands to Complete a Bracket

In this exercise you use an assortment of editing commands to complete a simple drawing of a bracket. You could complete the drawing by drawing the required objects, but in this case and most cases, it is quicker to use the modify commands.

Completing a Bracket

1.   Open the drawing *brack.dwg*. The drawing looks like the following figure:

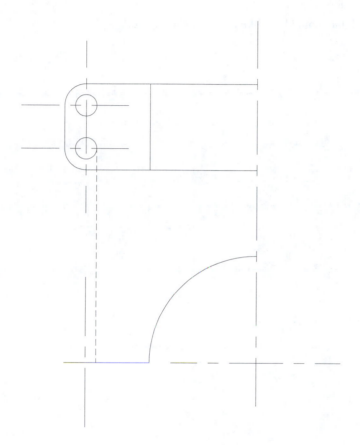

Figure 12-13: Brack.dwg

The drawing shows an incomplete plan and side view of a bracket. You need to complete the annotation lines and the outlines of the object. First you copy the hidden line to indicate the other part of the hole.

2. Start the COPY command and select the magenta dashed vertical line. This hidden linetype indicates the holes in the plan and side views.

Copy the line over the left side of the hole. To select the base point, use an intersection osnap and select the intersection of the hole and the red hidden line.

Specify the displacement by choosing the intersection of the hole and the red hidden line on the left side of the hole; you can use a quadrant or intersection object snap for this.

> Note You could also have entered a displacement of -0.5 in the Y direction.

The copied line should look like the following figure:

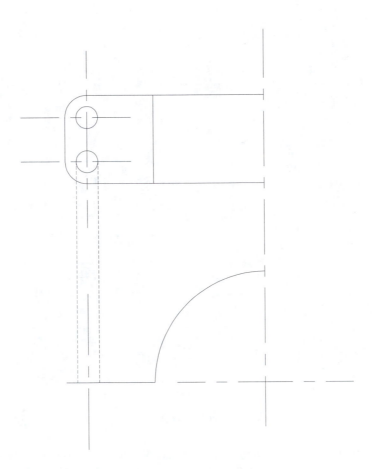

Figure 12-14: Copied line added to drawing

3. Now complete the side view of the bracket. Start the OFFSET command. Specify an Offset distance of **0.25**.

At the Select object to offset prompt, choose the polyline shown as L1 in the following figure:

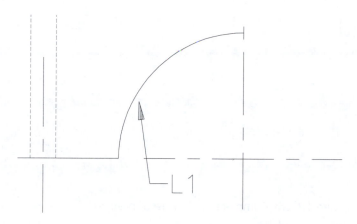

Figure 12-15: Polyline to choose

4. At the `Side to object` prompt, choose any point above the outline.

    Press ENTER at the `Select object to offset` prompt.

5. Draw a line to close the left end of the side view of the bracket.

    The bracket should look like the following figure:

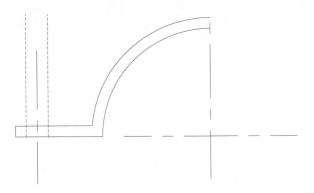

Figure 12-16: Completed side view of bracket

To complete the bracket, you need to mirror it around the central axis. It is easier to do this if you turn Ortho on.

6.   Start the MIRROR command and turn Ortho on.

At the `Select objects` prompt, enter **all** to select all objects in the drawing and then enter **r** to remove both the two center lines on the right side of the bracket.

For the `First point of mirror line` prompt, use an object snap to select one **end** of the main center line.

For the `Second point`, choose a point near the other end of the center line. You do not have to choose an exact point, because ortho forces the mirror to work in a horizontal or vertical direction only.

At the prompt `Delete old objects`, enter **No**.

The completed bracket should look like the following figure:

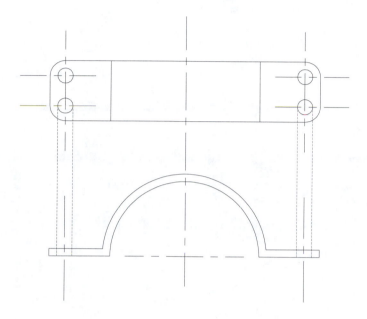

Figure 12-17: The completed bracket

This completes the exercise.

Notice that you did not need to worry about changing layers to get the correct linetypes while creating the drawing. Editing commands maintain the layers and properties of the edited objects.

Conclusion

After completing this chapter, you have learned the following

- ▶ There are several ways to create selection sets using various commands.

- ▶ The ERASE command removes selected objects from the drawing, and is probably the easiest editing command to use.

- ▶ Use the U command to undo a single action and the UNDO command to undo several actions.

- ▶ You can use the COPY, ARRAY, MIRROR, and OFFSET commands to create new geometry from existing objects.

Chapter 13

Modify Commands - Creating New Objects

In this chapter, you learn how to use AutoCAD Modify commands to create new objects using the ARRAY command.

About This Chapter

In this chapter, you will do the following:

▶ Create new objects from existing ones.

Using the ARRAY Command

The ARRAY command lets you create accurate and multiple copies of a selection set easily.

Methods for invoking the ARRAY command include:

- ◗ **Toolbar:** Modify
- ◗ **Menu:** Modify > Array
- ◗ **Command**: ARRAY

The prompts for the ARRAY command are:

```
Command: Array
Select objects: (Select object(s))
Select objects: (Press ENTER)
```

After selecting the object(s) to be arrayed, AutoCAD then requests you to indicate what type of array pattern you require:

```
Rectangular or Polar array (R/P) <R>:
```

The Rectangular option creates a rectangular array of rows and columns that form a matrix of the selected objects. Further prompts let you specify the number of rows and columns and the spacing between them.

```
Number of rows (---) <1>:
Number of columns (||||) <1>:
Distance between rows (---): (prompt if rows more than 1)
Distance between columns (||||): (prompt if columns more than 1)
```

The Polar option creates a circular array of the selected objects, copied around a central point. Further prompts let you set the center point, the amount of the rotation, (portion of the circle covered) and the orientation of the objects.

```
Center point of array:
Number of items:
Angle to fill (+=ccw, -+cw) <360>: (portion of circle - full circle is
default)
Rotate objects as they are copied? <Y> (keeps the objects orientated to
the center point)
```

Exercise 13-1: Using the Array Rectangular and Array Polar Commands

In this exercise, you use existing objects to create and position new drawing objects. You complete a simple drawing of a hub and then create multiple copies of the hub quickly and accurately with the ARRAY command.

Using the Array Rectangular and Array Polar Commands

1.   Open the file *hub.dwg*. The drawing looks like the following figure:

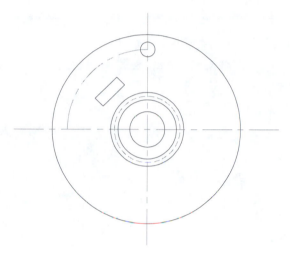

Figure 13-1: Hub.dwg

2.   First complete the outer ring of holes and the center lines. Start the ARRAY command.
Select the outer hole and center line, shown as P1 and P2 in the following figure:.

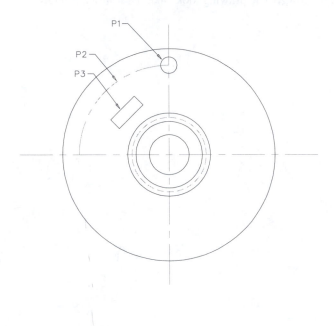

Figure 13-2: Objects to select

3. Use the polar option to create 4 instances of the objects, centered on the center of the hub. Make sure the objects are rotated as they are copied.

The completed prompts should look like this:

Command: **array**

Select objects: 1 found

Select objects: 1 found (select the circle and center line)

Select objects: (press Enter)

Rectangular or Polar array (<R>/P): **p**

Base/<Specify center point of array>: **cen** of (select any point on the circumferences of the circles sharing the common center)

Number of items: **4**

Angle to fill (+=ccw, -=cw) <360>: (press Enter)

Rotate objects as they are copied? <Y> (press Enter)

The hub should look like the following figure:

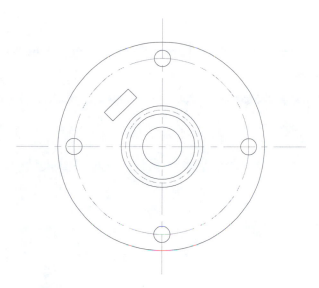

Figure 13-3: Hub after array

4. Now create a polar array of the slot, and create 8 slots in the hub. Make sure that you rotate the objects as they are copied.

The completed hub looks like the following figure:

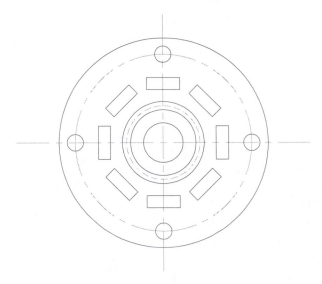

Figure 13-4: Hub with 8 slots

5.   Finally use the ARRAY command rectangular option to create 2 rows and 3 columns of the whole hub. Do not include the objects with center or hidden linetypes in the array. Specify a distance of 6 between the rows and 8 between the columns.

You will need to zoom to the extents of the drawing to see all the results. The complete drawing should look like the following figure.

Do not save your changes.

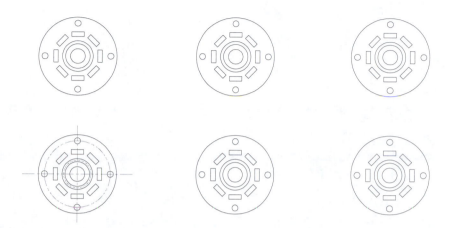

Figure 13-5: Hub after using ARRAY commands

Using the BREAK Command

The BREAK command erases parts of objects or splits an object in two.
Methods of invoking the BREAK command include:

▶ **Toolbar:** Modify > Break

▶ **Menu:** Modify > Break

▶ **Command:** BREAK

The Command prompts are:

Command: **break**

Select object: (select a point)

Enter second point (or F for first point): (Enter F or select another point)

To break off one end of a line, arc, or polyline, specify the second point beyond the end to be removed. To split an object in two without erasing a portion, enter the same point for both the first and second points. You can do this by entering @ to specify the second point. If you specify the second point, AutoCAD erases the portion between the first and second points. If the second point is not on the object, AutoCAD selects the nearest point on the object.

AutoCAD converts a circle to an arc by removing a piece starting counterclockwise from the first to the second point. The following figure shows examples of the BREAK command with circles and lines. The behavior of the command on a circle depends on whether you select point X or point Y first.

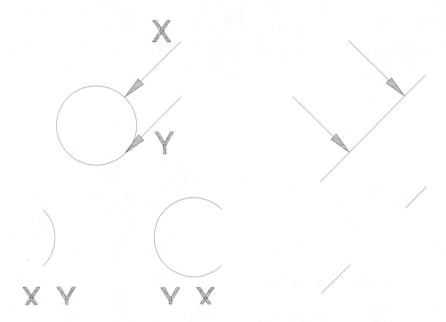

Figure 13-6: Examples of using the BREAK command

Using the EXPLODE Command

The EXPLODE command breaks a compound object into its component objects. Compound object are objects made up of more than one AutoCAD object, such as a multiline.

Methods for invoking the EXPLODE command.

▶ **Toolbar**: Modify > Explode

▶ **Menu:** Modify > Explode

▶ **Command:** EXPLODE

The command prompts you for objects to explode. Here is a summary of the results of exploding common drawing objects. For a complete list, see the online Help.

▶ *Block* - Blocks with equal X, Y, and Z scales explode into their component objects. Exploding a block with contains attributes deletes the attribute values and redisplays the attribute definitions

▶ *2D polyline* - Discards any associated width or tangent information

▶ *Polyline with width* - Places the resulting lines and arcs along the center of the polyline

▶ *Multiline* - Explodes into lines

> ◗ *Dimensions* - Explodes into mtext, lines, solids and points

> ◗ *Hatch Patterns* - Explodes into constituent lines

Note: The EXPLMODE system variable controls whether EXPLODE supports non-uniformly scaled blocks.

Using the TRIM and EXTEND Commands

Often it is easier to trim back an object, rather than having to redraw it. The TRIM command lets you quickly and accurately trim objects.

Methods for invoking the Trim command include the following:

> ◗ **Toolbar:** Modify > Trim

> ◗ **Menu:** Modify > Trim

> ◗ **Command:** TRIM

The TRIM command only works with arcs, elliptical arcs, lines, open 2D and 3D polylines, rays and splines.

```
Command: Trim
Select cutting edges: (Projmode = UCS, Edgemode = No extend)
Select object: (Select object(s))

Select object: (Press ENTER)
<Select object to trim>/Project/Edge/Undo: (Select objects to trim or
enter option)
```

At the `Select objects` prompt, you select the object(s) to define the cutting edges to which object(s) are trimmed. Then you select the object(s) to be trimmed. The EDGEMODE option, when enabled, trims the object(s) to its natural boundary, even if the object does not physically intersect the cutting edge. When disabled, TRIM only trims the object(s) if it actually intersects the boundary.

The PROJMODE variable specifies the Projection mode used when trimming objects. It is used for 3-dimensional situations where objects that have an apparent intersection can be used as a basis for trimming.

Because of changes to objects in the drawing you often need to be able to quickly and precisely adjust/extend part of the object(s) to another point/object. This is precisely what the EXTEND command does, however the command only works with arcs, elliptical arcs, lines, open 2D and 3D polylines, and rays.

Methods of invoking the EXTEND command include:

> ◗ **Toolbar:** Modify > Extend

> ◗ **Menu:** Modify > Extend

▶ **Command:** EXTEND

The following figure illustrates the function of the TRIM and EXTEND commands:

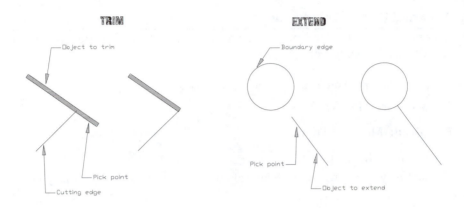

Figure 13-7: Comparison of the TRIM and EXTEND commands

Using the LENGTHEN Command

This command is based on the TRIM and EXTEND commands. The trim and extend operations used in LENGTHEN can be performed on existing objects. The LENGTHEN command options allow dynamic dragging to indicate the length, input of a numeric delta value, a percentage value, a change to the included angle of an arc, total length and lets you repeat the selection of objects.

▶ **Toolbar:** Modify > Lengthen

▶ **Menu:** Modify > Lengthen

▶ **Command:** LENGTHEN

LENGTHEN works fully with lines or arcs only. Some options work with elliptical arcs and two point polylines. The Command prompts are:

```
Command: Lengthen
DElta/Percent/Total/ DYnamic/<Select Object>:
```

▶ *Delta* - Specifies an incremental length or angle. The value can be entered by number or by selecting two points. A positive value extends, a negative value trims.

▶ *Percent* - Sets a new length or angle by specifying a percentage of its original length. Only positive numbers are allowed. For example, 75% trims off 25%

of original length from chosen endpoint.

▶ *Total* - You can specify a total length or angle (in degrees), then select the object's endpoint. The value can be entered by number or by selecting two points. Only positive numbers are allowed.

▶ *Dynamic* - Lets you drag a line, arc or elliptical arc by selecting one endpoint, while the other endpoint stays fixed.

Selecting an object indicates the current values for an object. The following figure shows two options of the LENGTHEN command used on an arc:

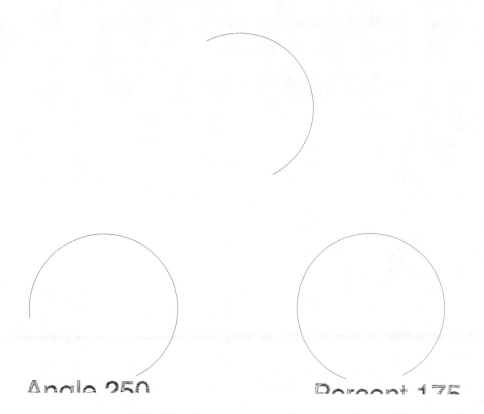

Angle 250   Percent 175

Figure 13-8: Changing an arc with the LENGTHEN command Total options

Using the FILLET and CHAMFER Commands

The FILLET command connects two lines, arcs, or circles with an arc. You specify the radius of the arc. The lengths of the original lines and/or arcs can be retained (depending on the method chosen) or changed by the FILLET command to smoothly (tangentially) fit to an arc. FILLET can also create a sharp corner at the intersection of two lines by specifying a radius of 0.

FILLET is used by designers to represent fillet welds and by architects to connect wall intersections.

The following figure shown examples of the FILLET command:

Figure 13-9: Examples of using FILLET

Fillets can be created between any combination of two lines, arcs, or circles. If more than one fillet can be created, then AutoCAD chooses the fillet whose endpoints are closest to the points you choose.

Methods for invoking the FILLET command include:

- ▶ **Toolbar:** Modify > Fillet

- ▶ **Menu:** Modify > Fillet

- ▶ **Command:** FILLET

The prompt sequence for the FILLET command is:

```
Command: fillet
(TRIM mode) Current fillet radius = 0.0000
```

Polyline/Radius/Trim/<Select first object>: (Select an object)
Select second object: (Select an object)

Other options include:

▶ *Polyline* - You select a Polyline all its vertices are filleted

▶ *Radius* - Sets the radius value for the fillet. Since FILLET has a default radius of 0, you must establish your radius first. A value entered for the fillet radius is stored with the drawing and remains the same until changed.

▶ *Trim* - Controls whether the edges used to form the fillet are retained or removed. Values are Trim/No Trim, the default is Trim.

You can make two lines intersect by starting the FILLET command and setting the fillet radius to 0, then selecting the two lines to effectively extend or trim them to an intersection point.

Closing or capping of parallel lines is a common drafting task in many disciplines. Using the FILLET command, the cap radius distance is automatically calculated between any two parallel lines, rays or xlines. It ignores the fillet radius at the end closest to where you select.

Note: The xline cannot be the first object selected.

The order in which you select the elements that make up the fillet makes no difference. The FILLET command is shown in the following figure:

Figure 13-10: Capping parallel lines with the FILLET command

Using the CHAMFER Command

Chamfers are used in mechanical applications, such as beveled edges on turned shafts, to eliminate sharp corners or rough edges. Chamfers make it easier to assemble parts. They are also used to ease edges on furniture and poured in-place, structural concrete columns.

The CHAMFER command trims two intersecting lines a specified distance from their intersection, and connects the trimmed ends with a beveled edge. Its operation is similar to the FILLET command. CHAMFER works even when the two lines do not intersect. Such lines are extended to their intersection point, and are then trimmed back and beveled.

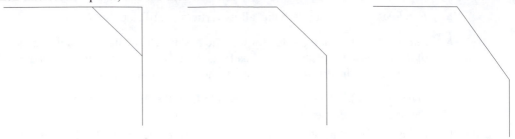

Figure 13-11: Examples of using CHAMFER

Methods for invoking the CHAMFER command include:

- **Toolbar:** Modify > Chamfer

- **Menu:** Modify > Chamfer

- **Command:** CHAMFER

The prompt sequence for the CHAMFER command is:

```
Command: chamfer
(TRIM mode) Current chamfer Dist1 = 10.0000, Dist2 = 10.0000
Polyline/Distance/Angle/Trim/Method/<Select first line>: (Select first
line)
Select second line: (Select second line)
```

Other options include:

- *Polyline* - You can select a Polyline. All vertices are chamfered.

- *Distance* - The distance to trim back from the first line selected, and the distance to trim the second line (they do not have to be the same).

Note: Always establish the distances for a chamfer first.

- *Angle* - Chamfer angle, measured from the first line

- *Trim* - Controls whether the edges used to form the chamfer are retained or removed. Values are Trim or No Trim; the default is Trim.

- *Method* - Sets the chamfer creation method. Two distances or distance and angle. You use this option in the following exercise.

You can use the CHAMFER command with splines, ellipse, 3D polylines, rays and xlines as well as segments of polylines, polyline arcs, and lines.

> Note: You can use the TRIMMODE system variable to set whether you retain the original geometry with a fillet or chamfer. Setting TRIMMODE to 0 retains all the original geometry, setting it to 1 discards the old geometry.

◗ The FILLETRAD variable contains the current fillet radius setting.

◗ CHAMFERA and CHAMFERB contain the two chamfer distances when used in the two distance mode.

◗ CHAMFERC sets the chamfer distance when used in association with CHAMFERD, the chamfer angle.

◗ Chamfers are created on the same layer as the objects if both the objects are on the same layer. They are created on the current layer if the source objects are on different layers.

Exercise 13-2: Using the Trim, Extend, Fillet, and Chamfer Commands

In this exercise you edit a drawing of a plate and create chamfers and fillets. You also modify existing objects by trimming and extending objects.

Using the Trim, Extend, Fillet, and Chamfer Commands

1. Open the file *plate.dwg*. The drawing looks like the following figure:

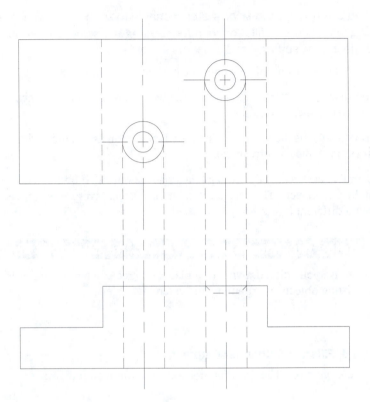

Figure 13-12: The plate drawing

2.   Methods for invoking the FILLET command include:

> ◗   **Toolbar:** Modify > Fillet
>
> ◗   **Menu:** Modify > Fillet
>
> ◗   **Command:** FILLET

3.   Set the fillet radius to **0.5** as follows:

Command: **fillet**

(TRIM mode) Current fillet radius = 1.0000

Polyline/Radius/Trim/<Select first object>: **r**

Enter fillet radius <1.0000>: **0.5**

4. Make sure that the command is in Trim mode. In the example above, Trim mode is set. If not, then restart the command, and enter **T**. Set the mode to **TRIM**, then press ENTER twice.

5. Start the command again, and choose the two lines that form the corner shown in the following figure:

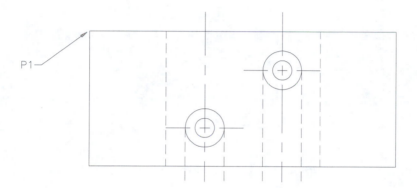

Figure 13-13: Objects to select for trimming

The filleted corner looks like the following figure:

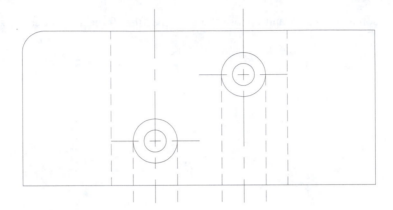

Figure 13-14: Filleted corner

6. Repeat the fillet for the other top corner, but this time set the trimming mode to NOTRIM by choosing the Trim option. Then specify the two edges to fillet as follows:

Command: **fillet**

(TRIM mode) Current fillet radius = 0.5000

Polyline/Radius/Trim/<Select first object>: **t**

Trim/No trim <Trim>: **n**

Polyline/Radius/Trim/<Select first object>: (Select one edge to fillet)

Select second object: (Select other line making up the corner)

This time the original edges are retained. Now you will chamfer the lower corners.

7. Enter **chamfer** at the command prompt to start the CHAMFER command.

8. You will set a first chamfer distance that is different from the second. Set the chamfer distances by choosing the Distance option and setting the first distance to **0.25**, the second to **0.75**.

Command: **chamfer**

(NOTRIM mode) Current chamfer Dist1 = 0.5000, Dist2 = 0.5000

Polyline/Distance/Angle/Trim/Method/<Select first line>: **d**

Enter first chamfer distance <0.5000>: **0.25**

Enter second chamfer distance <0.2500>: **0.75**

Now you select the edges to chamfer. In this case, the order in which you choose does make a difference.

9.   Start the CHAMFER command, and at the prompts, choose P1, then choose P2.

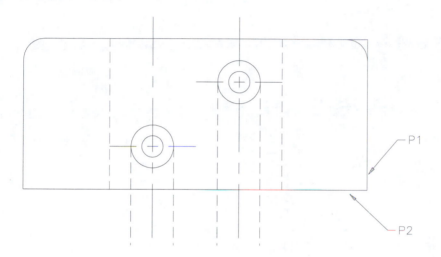

Figure 13-15: Completed chamfer

You are still in NOTRIM mode, so the original edges remain.

10.   Change to TRIM mode, and chamfer the remaining edge of the plate.

Using Chamfer with Distance and Angle

1.   You can also chamfer edges based on a distance and angle method. Start the CHAMFER command and change to the angle method by entering **method**, and then entering **angle**.

You need to specify angle twice. Then enter a chamfer length of **0.75** on the first line, and an angle of **15** from the same line.

Command: **CHAMFER**

(TRIM mode) Current chamfer Dist1 = 0.2500, Dist2 = 0.7500

Polyline/Distance/Angle/Trim/Method/<Select first line>: **m**

Distance/Angle <Distance>: **a**

Polyline/Distance/Angle/Trim/Method/<Select first line>: **a**

Enter chamfer length on the first line <1.0000>: **0.75**

Enter chamfer angle from the first line <0>: **15**

2. Now you can choose the objects to chamfer. Start the CHAMFER command, and choose P1 and P2 in the following figure for the first and second lines:

The result of the CHAMFER command is shown in the following figure:

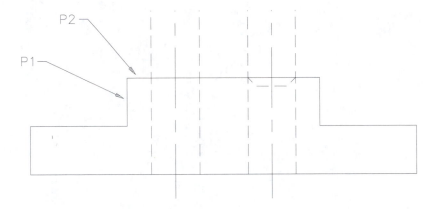

Figure 13-16: Chamfered side view

3. Repeat the chamfer for the other edge of the side view.

The hidden lines between the two views are not required. You can discard the parts you do not need with the TRIM command.

4. From the Modify menu, choose Trim.

5. At the `Select cutting edges` prompt, select the line shown by P1 in the following figure, then press ENTER.

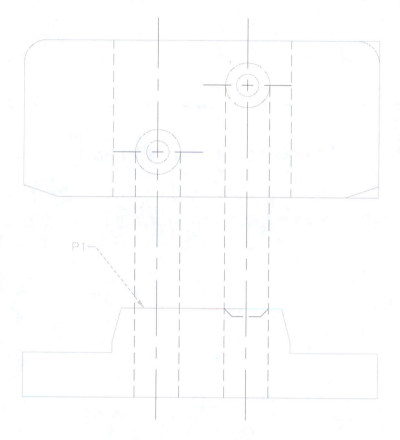

Figure 13-17: Line to select for chamfer

To save time, you can extend this line so it will act as a trimming edge all the way across the drawing.

6.   Set EDGEMODE as follows:

<Select object to trim>/Project/Edge/Undo: **e**

Extend/No extend <No extend>: **e**

Notice that the highlighted cutting edge goes completely across the top of the side view.

7.   Complete the operation by choosing the hidden lines above the highlighted edge.

The completed trim looks like the following figure:

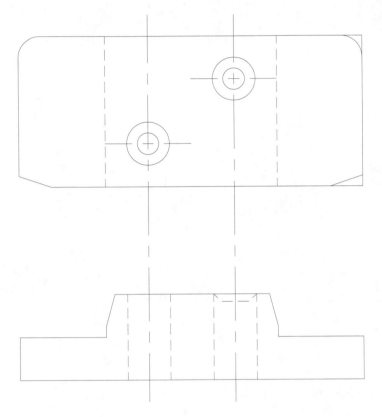

Figure 13-18: Completed trim

8. Finally complete the countersunk holes in the side view of the plate by copying the countersink in the right hole of the side view to the appropriate location on the left side.

The completed plate looks like the following figure:

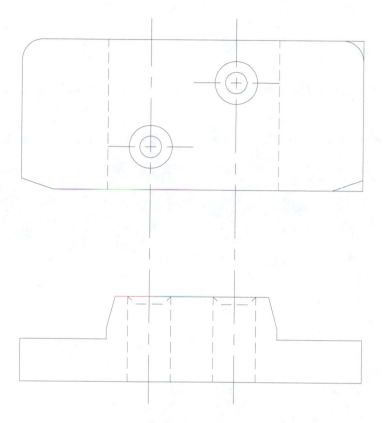

Figure 13-19: The completed plate

Conclusion

After completing this chapter, you have learned the following

▶ The ALIGN, MOVE, and ROTATE commands are used to change the location of existing geometry.

▶ The STRETCH, SCALE, FILLET, BREAK, LENGTHEN, and CHAMFER commands can be used to change object forms and finish shapes.

Chapter 14

Polyline Edit and Grips

In this chapter, you will learn how to use the PEDIT command as well as how to edit objects using Grips.

About This Chapter

In this chapter, you will do the following:

▶ Edit polylines and splines.

▶ Use grips to quickly edit objects.

Using the PEDIT Command

One way to edit a polyline is with the PEDIT command. Methods for invoking the PEDIT command include:

▶ **Menu:** Modify > Object > Polyline

▶ **Command:** PEDIT

After you select a polyline, this prompt is displayed:

Close/Join/Width/Edit vertex/Fit/Spline/Decurve/Ltype gen/Undo/eXit <X>:

Notice you can only edit one polyline at a time, although you can use the Join option to join two or more polylines into one polyline. Only polylines can only be joined at a single vertex point provided that they do not continue through the vertex point.

You can use grips to edit polyline vertices. You can also use DDMODIFY to change some properties of a polyline, and edit a polyline with commands like CHPROP, TRIM, OFFSET and MIRROR.

If you use PEDIT or DDMODIFY to change the polyline width, the whole polyline is changed. You cannot alter the width of individual segments with these commands. If you select the Edit vertex option, an X is displayed at the first vertex of the polyline, and this prompt is displayed:

Next/Previous/Break/Insert/Move/Regen/Straighten/Tangent/
Width/eXit <N>:

You will use some of these options to modify polyline vertices in the following exercise.

You can use the PEDIT command to create a curve through a series of vertices in a polyline. The nature of the curve depends on which option you use. *Fit curves* are curves that pass through all the vertices of a polyline. The curve consists of a pair of arcs joining each pair of vertices: extra vertices are added to the polyline to enable this. A *spline curve* uses the vertices of a polyline as control points, or a *frame*, of a curve. The curve passes through the first and last control points, and is pulled towards the other control points. This type of curve is known as a *B-spline*. In general, splines create a smooth curve with only a few control points.

> Note: Using the Spline option does not create a spline object (which is created by the SPLINE command), but instead creates a spline fit polyline.

The SPLINETYPE system variable controls the type of spline curve generated, while the SPLINESEGS variable controls the coarseness of the curve.

These curves can be returned to their original forms with the Decurve option of the PEDIT command. You can see the frame of a spline curve by setting SPLFRAME to 1. To edit a spline with grips, show the frame and edit the frame points only. If you enter the spline curve itself, your results are lost when you exit the PEDIT command.

The DDMODIFY command lets you select an object, and then shows its properties in a dialog box. In this dialog box you can view and edit some of the object properties. For a polyline, you can change these properties:

▶ The curve type and fitting method

▶ Whether a polyline is closed or open

▶ How linetype patterns are applied

Note: You can only use DDMODIFY with one object at a time. The appearance of the dialog box changes based on the object you select.

Using the Optimized Polylines

Polylines are one of the most frequently used objects in AutoCAD. You can use polylines to represent walls, boundaries, edges, or other geometry as one drawing object. You can assign line widths to polylines as well. However, the drawing overhead associated with a polyline is relatively high. In Release 14, most polylines are automatically converted to optimized polylines. The optimized polyline provides most of the functionality of 2D polylines with much improved performance and reduced database (and hence drawing file) size.

Optimized polylines have no special commands associated with them. They are created with the PLINE command and edited with the PEDIT command.

The optimized polyline supports all polyline features, except

▶ Spline fitting

▶ Curve fitting

When you use the PEDIT command to create these polylines, the command transparently changes the polyline to a 2D polyline.

Using the Convert Command for Polylines

With the CONVERT command, you can manually convert any 2D polylines or hatch patterns to optimized polylines and Release 14 hatch objects respectively. The CONVERT command converts Release 13 associative hatched areas to the new Release 14 hatch object with the exception of Solid filled hatches. Solid filled hatches are converted to non-associative blocks composed of 2D solid objects. The command is generalized to act on all objects that have replacements in Release14 (presently 2D polylines and hatch objects).

The command uses the Command line interface for converting objects. When you enter **convert** at the Command prompt, you are asked for which object type to convert (hatch objects, polyline objects, or both hatch and polylines). Finally, you select the objects to convert. Any selection method is valid, and the default is All. If you manually select objects, any object that is not valid is filtered out of the selection set. The Command prompt output is displayed as follows:

```
Command: convert
Specify object type.  Hatch/Polylines/<Both>:
Select objects to convert <All>:
```

Editing Polylines with Grips

Grips are a handy way to edit polylines. Changing the control points of one of these curves is quicker ands simpler than using the PEDIT command.

> Note: The SPLINETYPE system variable controls the type of spline curve generated, while the SPLINESEGS variable controls the coarseness of the curve.

Editing Objects Using Grips

This section provides an overview of the operation and uses of object grips. Object Grips (referred to as grips) let you edit AutoCAD drawings in an entirely different way. Without even entering an AutoCAD command, you can stretch, move, copy, rotate, scale, and mirror objects. You can also automatically snap to geometric features of objects such as endpoints, midpoints, quadrants, and centers with grip editing options. There are also rectangular and circular auxiliary snap grids available within grip editing.

In this section you learn how to manipulate objects with hot grips, warm grips, and cold grips. You also learn how to take advantage of grips for feature-based snapping.

Grips Dialog Box

To use grips, select the Grips option from the Tools pull-down menu to display the Grips dialog box, and then check the Enable Grips check box.

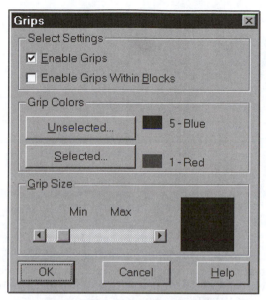

Figure 14-1: The Grips dialog box

AutoCAD tells you when grips are enabled by displaying a *pickbox* at the intersection of the crosshairs. The pickbox also is displayed when Noun/Verb Selection (the PICKFIRST system variable) is turned on in the Selection Settings dialog box. To display the crosshairs cursor without a pickbox, both GRIPS and PICKFIRST must be turned off.

Grip States

Once grips are enabled, squares are displayed on objects when you pick them without starting a command.

A *hot grip* is a grip you select with the cursor. It has a solid filled color, and is the grip that will be operated on. A *warm grip* is a grip that you have not selected with the cursor but is on an item in the current selection set. You can tell a grip is warm because the object is highlighted and the Command prompt shows the Stretch option. Objects with warm grips are affected by grip command options on a hot grip (with the exception of the Stretch option). A *cold grip* is a grip on an object that is not in the current selection set but you can still use it to snap to. Notice objects with cold grips are not highlighted.

The following figure shows (from left to right) hot, warm, and cold grips:

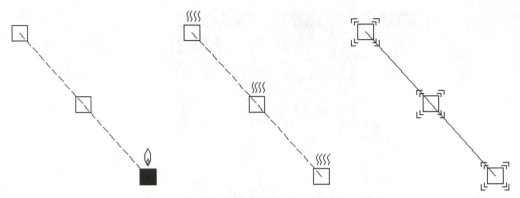

Figure 14-2: Hot, Warm, and Cold Grips

Selecting an object makes the grips warm. Removing an object from the current selection set changes the grips from warm to cold. You do this by pressing the SHIFT key while re-selecting the object. Selecting a warm grip makes it hot. Pressing ESC (Escape) cancels grip operations by clearing the selection set. A second ESC removes cold grips. To make more than one object hot for an operation, hold down the SHIFT key as you select warm grips.

Grip Editing Methods (Modes)

When you select a warm object grip, this Command prompt is displayed:

```
** STRETCH **
<<Stretch to point>>/Base point/Copy/Undo/eXit:
```

At this prompt, you can select a new point in the drawing. The grip you selected is relocated to the selected point, stretching the rest of the object associated with that grip. The Stretch option is shown in the following figure:

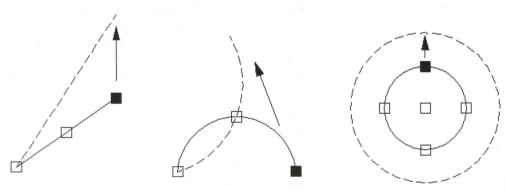

Figure 14-3: The Stretch option

You can cycle through the other grip edit options by pressing ENTER, the space bar, or the equivalent button on your puck or mouse.

```
** MOVE **
<<Move to point>>/Base point/Copy/Undo/eXit:
** ROTATE **
<<Rotation angle>>/Base point/Copy/Undo/Reference/eXit:
** SCALE **
<<Scale factor>>/Base point/Copy/Undo/Reference/eXit:
** MIRROR **
<<Second point>>/Base point/Copy/Undo/eXit:
```

You can also choose the options by right-clicking the mouse when a grip is hot.

Stretch

You are not just limited to stretching objects with grips. You can also use the sub-options of the Stretch option to define a base point for the stretch or make copies of the original object as you stretch it. For example, you could use the Copy sub-option to create many lines radiating from an opposite endpoint, as shown in the following figure. Points P1, P2 and P3 represent the points you select for the stretch point.

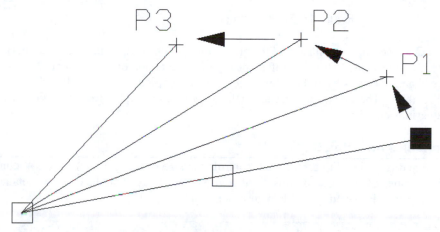

Figure 14-4: Stretching with the Copy sub-option

Grips are a handy way to edit spline and fit polylines. Changing the control points of one of these curves is quicker ands simpler than using the PEDIT command.
The following two items are examples of using the grip sub-options to rapidly edit a drawing:

> ▶ Use the ROTATE and Copy sub-options to form a polar array of selected objects.

> ▶ Use the SCALE and Copy sub-options to create a series of concentric rings from a circle.

Snapping to Grip Locations

As you move the cursor around a drawing with cold, warm, or hot grips, you'll notice that it snaps or locks onto an object grip when the cursor moves into the square zone representing the grip.

You can use this instead of using an object snap such as ENDpoint or MIDpoint.

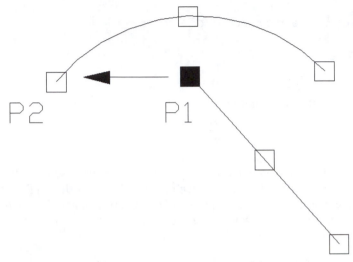

Figure 14-5: Snapping to a Grip

If you use the Copy sub-option with any of the grip editing commands, a temporary auxiliary snap grid is created if you use the SHIFT key after specifying the location of the first copy. AutoCAD then uses the X and Y offsets from the original object to the copied object to define the snap grid spacing and rotation. The snap grid is rectangular except when using the Rotate option, in which case the snap grid is circular.

System Variables Affecting Grips

GRIPS turns on and off grip editing. GRIPSIZE controls the size of the grips. GRIPCOLOR controls the color of the grips. GRIPBLOCK controls whether grips are displayed on the objects contained in a block, or one grip is displayed at the block insertion point.

Exercise 14-1: Editing With Grips

In this exercise, you become familiar with using grips to modify an existing drawing of an office. You use grips to rotate, move, and copy existing objects.

Using Grips

1.   Open the file, *g-office.dwg*. The drawing looks like the following figure:

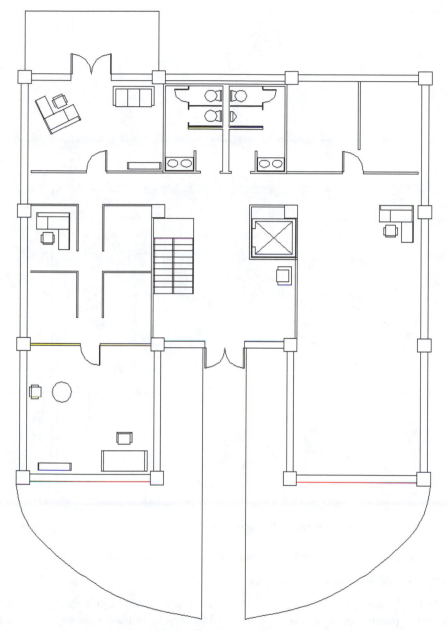

Figure 14-6: G-office.dwg

2. This is a drawing of an office layout. You will make some changes to the furniture layout using grips. Choose the Zoom Window button and zoom in on the upper left corner of the building, as shown in the following figure:

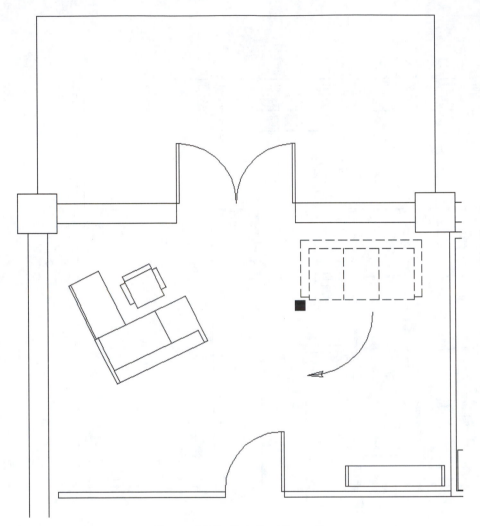

Figure 14-7: Rotating the sofa

3.   Now move and rotate the sofa into a new position. Select the sofa.

Only one grip is displayed because the sofa is a Block.

4.   Select the warm grip (the one grip on the sofa) to make it hot. Cycle through the options with the spacebar or right mouse button until the Rotate option is displayed. Rotate the sofa **-90** degrees.

5.   Now move the sofa against the right wall of the office and then press ESC twice to remove the grip from the sofa.

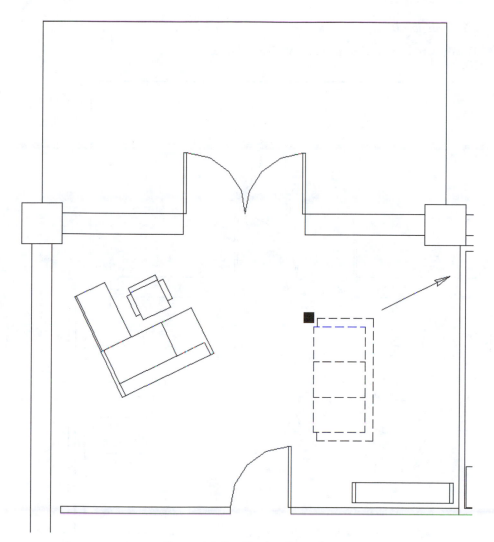

Figure 14-8: Moving the sofa

6. Next you will rotate the desk so that it is at 90 degrees using the Reference sub-option. Points P1 and P2 in the following figure specify the reference angle. You can easily snap to P1 because of the grip there. You will need to use the ENDpoint object snap to select P2.

Select the desk and complete the prompts as follows:

Command: (Select the grip shown at P1 in Figure 11)
ROTATE
<Rotation angle>/Base point/Copy/Undo/Reference/eXit: **R**
Reference angle <0>: (Select P1 again)

Second point: **end** *of* (Select P2)

<New angle>/Base point/Copy/Undo/Reference/eXit: **90**

The desk rotation is shown in the following figure:

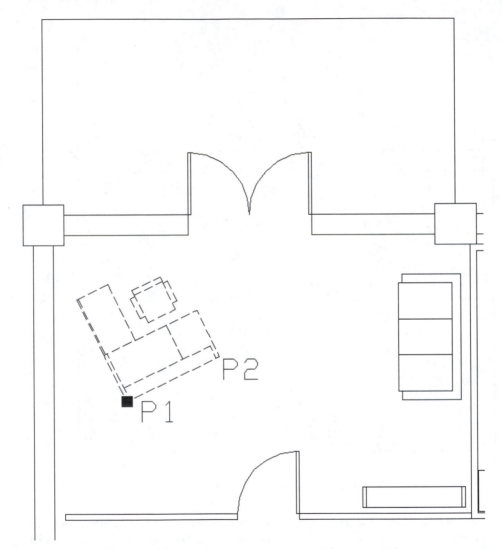

Figure 14-9: Rotate the desk using a reference angle

7.   Move the desk into its final position with the Stretch or the Move options.

Since the desk is a block, these options behave the same way.

The position of the desk is shown in the following figure:

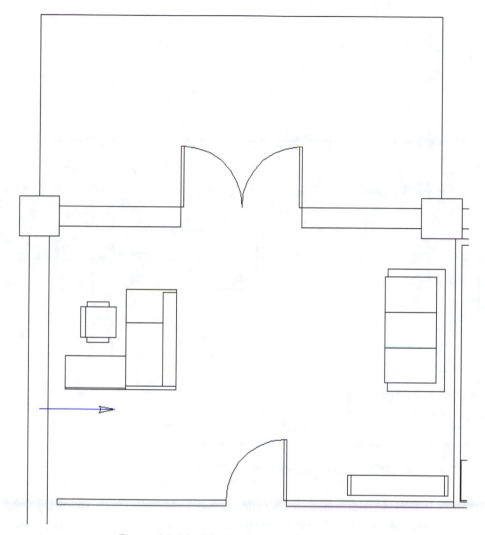

Figure 14-10: Moving the desk into position

8.   ZOOM into the lower left corner of the office building. Select the executive desk in the lower right corner of the office.

9.   Enable Ortho using the F8 key or by double clicking the Ortho button on the status bar.

10.   Use the grip feature to mirror the desk with both the Copy and the Base point sub-options. Locate the base point at the center of the chair.

Your results should look like the following figure:

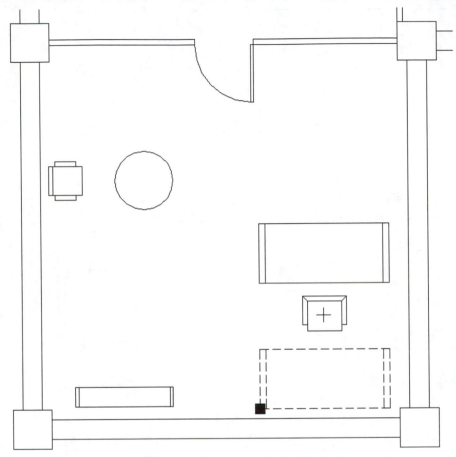

Figure 14-11: Mirror and copy the desk with the Base option

11. Continue working with grips to make your drawing look similar to the following figure. Do not forget to use the Copy option when appropriate.

When making multiple copies at a regular spacing, try holding down the SHIFT key after the first copy.

If you need to manipulate more than one Block, hold the SHIFT key when selecting hot grips. You can select and manipulate both with the same grip editing option.

The completed drawing is shown in the following figure:

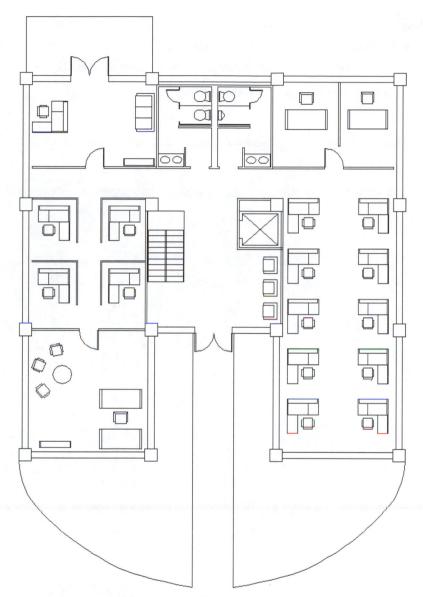

Figure 14-12: Completed drawing

Conclusion

After completing this chapter, you have learned the following:

▶ The PEDIT command is used to edit polylines to create smoothed curves.

▶ Grips can carry out many editing operations.

Chapter 15

Inquiry Commands

In this chapter, you learn how several AutoCAD commands let you display a variety of information about your drawing and the objects drawn within. With the use of inquiry commands, objects that are drawn to full size can be used to provide information such as the area of a room from an architectural floor plan drawing to determine how much carpet to order, or the bounding box of a common gasket to maximize the cut layout on a sheet of gasket material. Inquiry commands can also provide database information for selected objects, display the status of a drawing, and document the time that you can spend working on the drawing.

About This Chapter

In this chapter, you will do the following:

▶ Display statistics about a drawing

▶ Locate the XYZ coordinates of a point location on the drawing

▶ View information about a specific object

▶ Determine the total distance, angle, and delta X,Y and Z distances between two points

▶ Obtain area and perimeter

▶ Check for elapsed drawing time

Using the STATUS Command

The STATUS command displays drawing statistics, modes, and extents in the Command History area of the AutoCAD Text window and of the Command window. All coordinates and distances are displayed by STATUS in the format specified by the UNITS or DDUNITS commands.

Methods for invoking the STATUS command include:

▶ **Menu:** Tools > Inquiry > Status

▶ **Command:** STATUS

The following figure shows the AutoCAD Text Window that opens upon accessing the STATUS command:

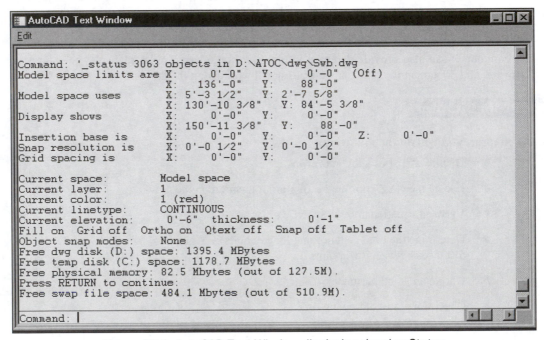

Figure 15-1: AutoCAD Text Window displaying drawing Status

The AutoCAD Text Window first reports the number of objects in the current drawing. This includes graphic objects such as arcs and polylines, non-graphic objects such as layers and linetypes, and internal program objects such as symbol tables. In addition, STATUS returns the following information:

▶ Model or Paper space limits   ▶ Current linetype

▶ Model or Paper space uses   ▶ Current elevation

▶ Display shows   ▶ Thickness

▶ Insertion base

▶ Snap resolution

▶ Grid spacing

▶ Current space

▶ Current layer

▶ Current color

▶ Fill, Grid, Ortho, Qtext, Snap, Tablet

▶ Object Snap modes

▶ Free *.dwg* disk space

▶ Free physical memory

▶ Free swap file space

Note: Entering Status at the DIM Command prompt displays the values and descriptions of all dimensioning system variables.

Displaying the XYZ Coordinate of a Point

The ID command is used to display the absolute XYZ coordinate values of a selected location, referencing the current UCS. Methods for invoking the ID command include:

▶ **Toolbar:** Inquiry

▶ **Menu:** Tools > Inquiry > ID Point

▶ **Command:** ID

Command prompt input and display of the ID command results should be:

```
Command: id
Point: _endp of  X = 77'-10"     Y = 68'-6"     Z = 0'-0"
```

After selecting the endpoint of a line, AutoCAD displays the UCS coordinate values of the location at the Command prompt.

Note: ID defines the specified point as the last point. You can reference the last point by entering @ at the next prompt that requests a point.

Displaying an Object's Database Information

The LIST command is used to display database information for the selected object or objects.

Methods for invoking the LIST command include:

▶ **Toolbar:** Inquiry

▶ **Menu:** Tools > Inquiry > List

▶ **Command:** LIST

The following figure shows the AutoCAD Text window displaying the LIST command information that was returned for a selected line:

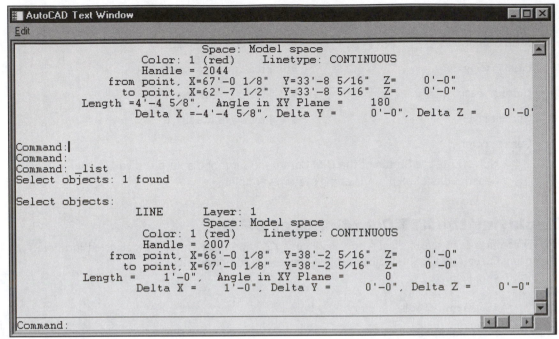

Figure 15-2: LIST command data for a line

AutoCAD lists the object type, object layer, and status of the object, whether it is in model space or paper space. LIST also reports color and linetype information if that information is not set to BYLAYER.

For the line selected in the previous figure, the from point and to point data defines the endpoint XYZ absolute coordinate locations. The Length is defined as the true 3D length of the line. The Angle in XY Plane is the direction the line was drawn, from point - to point. Delta X, Delta Y, Delta Z display the view length, parallel to the respective axis.

Note: The thickness of an object is displayed if it is non-zero.

Determining the Distance between Points

The DIST command measures the actual and delta X,Y and Z distances. DIST also returns the angle between the two points, relative to the current UCS.

Methods for invoking the DIST command include:

> **Toolbar:** Inquiry

> **Menu:** Tools > Inquiry > Distance

> **Command:** DIST

Command prompt input and display of the DIST command results should be as follows:

```
Command: '_dist
First point: _endp of  select an endpoint
Second point: _endp of select an endpoint
Distance = 7'-1 1/4",  Angle in XY Plane = 39,  Angle from XY Plane = 0
Delta X = 5'-6",  Delta Y = 4'-6",   Delta Z = 0'-0"
```

In the command history window, AutoCAD displays the true 3D distance between the points selected. The angle in the XY plane is relative to the current X axis. The angle from the XY plane is relative to the current XY plane.

Note: The distance unit values are displayed using the current units format.

Calculating a Bounded Area

With the AREA command, you can calculate the area and perimeter of objects or areas defined by a sequence of points. The combined area of more than one object can be calculated by keeping a running total as you add or subtract areas from the selection set.

Methods for invoking the AREA command include:

▶ **Toolbar:** Inquiry

▶ **Menu:** Tools > Inquiry > Area

▶ **Command:** AREA

Calculating a Defined Area

The following shows how AREA is used to return the area and perimeter of the rectangle by selecting the endpoints at ABCD. All points must lie in a plane parallel to the XY plane of the current UCS.

```
Command: _area
<First point>/Object/Add/Subtract: _endp of  Selected at A
Next point: _endp of Selected at B
Next point: _endp of Selected at C
Next point: _endp of Selected at D
Next point: ENTER
Area = 24.0000, Perimeter = 20.0000
```

The area of the rectangle looks like the following figure:

Figure 15-3: Finding the area of a rectangle

Calculate the Area Enclosed by an Object

Because the rectangle in the previous figure is a closed polyline, the area and perimeter can be displayed by selecting the object, as shown in the following example.

```
Command: _area
<First point>/Object/Add/Subtract: O for Object selection
Select objects:  (Select anywhere on the rectangle )
Area = 24.0000, Length = 20.0000
```

Note: The Perimeter value is replaced with a Length on a polyline object.

The Object option in the AREA command returns the enclosed area of an arc, circle, ellipse, lightweight polyline, polyline, region or planar closed spline. With open objects such as arcs, spline curves and open polylines, the area is computed as though a straight line connects the starting point and endpoint. The area defined by the center of the width is returned for wide polylines.

Calculate Combined Areas

With the AREA command it is possible to calculate the combined areas of selected objects or point locations. By entering the Add mode in the AREA command, you begin a running tally for the area bound by the points or objects selected. As long as you continue to select points or objects, AutoCAD will add up the total combined area.

The following input is used to define the combined area of the rectangle and circle in the following figure.

```
Command: _area
<First point>/Object/Add/Subtract: A Turns on Add mode and keeps a
running balance of additions
<First point>/Object/Subtract: O
(ADD mode) Select objects:  Select the rectangle
Area = 24.0000, Perimeter = 20.0000  Values for the rectangle
Total area = 24.0000  For rectangle
(ADD mode) Select objects:  Select the circle
Area = 12.5664, Circumference = 12.5664  Values for the circle
Total area = 36.5664  For both objects
(ADD mode) Select objects: ENTER
<First point>/Object/Subtract:  ENTER
```

The result of the combined area of the two shapes is shown in the following figure:

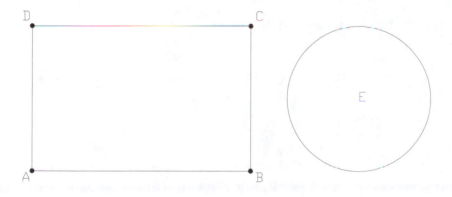

Figure 15-4: Combined area of 2 objects

Subtract Areas from Combined Areas

The AREA command can also be used to return the area of an object with internal objects removed from the overall area. The overall area is first defined with the Add mode. This value is needed to subtract from the internal areas. You then select the Subtract mode and begin to define points or objects for the area to subtract from the overall area.

The following command input is used to display the area of a rectangle with the area of the holes to be drilled removed from the overall area.

```
Command: _area
```

<First point>/Object/Add/Subtract: **A**
<First point>/Object/Subtract: **O**
(ADD mode) Select objects:
Area = 24.0000, Length = 20.0000
Total area = 24.0000
(ADD mode) Select objects: ENTER
<First point>/Object/Subtract: **S** Turns on Subtract mode, which works
the same as Add mode, except that it subtracts areas and perimeters.
<First point>/Object/Add: **O**
(SUBTRACT mode) Select objects: Select circle E
Area = 3.1416, Circumference = 6.2832
Total area = 20.8584
(SUBTRACT mode) Select objects: Select circle F
Area = 0.7854, Circumference = 3.1416
Total area = 20.0730 Of the rectangle with the 2 holes removed
(SUBTRACT mode) Select objects: ENTER
<First point>/Object/Add: ENTER

The following figure displays the command input results:

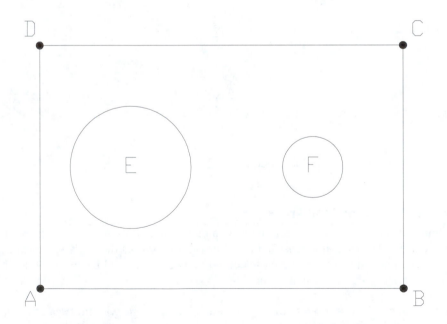

Figure 15-5: Rectangle with holes to subtract

Displaying Time Statistics

The TIME command displays the date and time statistics of a drawing.

Methods for invoking the TIME command include:

    ▶ **Menu:** Tools > Inquiry > Time

    ▶ **Command:** TIME

When you access the TIME command, the AutoCAD Text Window displays the time of original creation and latest revision. It displays the total editing time and the elapsed time for the current drawing session. TIME also displays the current time and the time of the next automatic saving of the drawing file.

TIME displays the following information about the current drawing:

```
Command: '_time
Current time:  Monday, April 14, 1997 at 7:13:40:950 PM
Times for this drawing:
Created:  Friday, June 16, 1995 at 10:20:50:740 AM
Last updated:  Monday, April 14, 1997 at 5:57:00:990 PM
Total editing time:  1 days 00:42:15.580
Elapsed timer (on):  1 days 00:42:15.580
Next automatic save in:  <no modifications yet>
Display/ON/OFF/Reset:  ENTER
```

The Display option redisplays the TIME data report. TIME can be turned on and off for the current drawing, and the timer can also be Reset. The timer feature is on by default.

Listing System Variables

AutoCAD stores the settings for the operating environment and some of the commands as system variables. Each system variable has an associated type: interger, real point, switch, or text string. The SETVAR command is used to list or change the values of system variables that are not read-only.

Methods for invoking the SETVAR command include:

    ▶ **Menu:** Tools > Inquiry > Setvar

    ▶ **Command:** SETVAR

When you select the SETVAR command you are first prompted for the varible name you would like to control. By entering a system varible name, you are prompted to change its value. Entering **?** at the setvar Command prompt, lets you see the current settings for a single variable or a defined list.

```
Command: '_setvar Variable name or ?: ?  Used to query a list
Variable(s) to list <*>: M*  Display all system variables that begin
with the letter M
MAXACTVP          16
MAXSORT           200
MEASUREINIT       0
MEASUREMENT       0
MENUCTL           1
MENUECHO          0
MENUNAME          " C:\PROGRAM FILES\AUTOCAD R14\support\acad"      (read
only)
MIRRTEXT          1
MODEMACRO         ""
MTEXTED           "Internal"
```

You can enter a specific system variable name in place of the question mark. The value for that system variable will then be displayed and the option of entering a new value is given.

> Note: You can also change the value of system variables at the Command prompt by entering the name of the variable and its new value without using the SETVAR command.

Exercise 15-1: Using Inquiry Commands

Inquiry commands let you display a variety of information about your drawing and the objects in the drawing. In this exercise you use Inquiry commands to display information about a fire sprinkler system drawing.

Using Inquiry Commands

1. Open the file *swb.dwg*. The drawing looks like figure 15-6. From the Tools menu, choose Inquiry, then choose Time. Press R and the ENTER to reset the timer. Press ENTER to complete the TIME command.

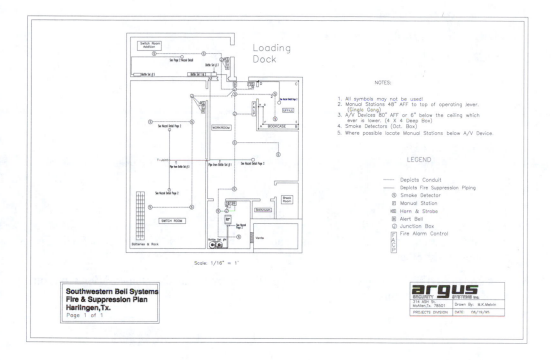

Figure 15-6: Swb.dwg

2. From the Tools menu, choose Inquiry and then choose Status. The AutoCAD Text Window is displayed, showing the drawing statistics, as shown in Figure 10-1. Notice especially the Free dwg disk space, Free physical memory, and Free swap file space settings. These are valuable system settings to know. Press ENTER to see the complete status list and complete the command.

3. Right-click on any toolbar button. The Toolbars dialog box is displayed. Check the Inquiry checkbox and then close the Toolbars dialog box.

4. From the Inquiry toolbar, choose the Locate Point button, which accesses the ID command. Select the lower left intersection of the drawing border. Remember to use Object Snaps.

   The coordinate values are displayed at the Command prompt. This is the absolute coordinate location of the lower left corner of the drawing extents, as follows:

   ```
   Point:   X = 5'-3 1/2"      Y = 2'-7 5/8"      Z = 0'-0"
   ```

5. Choose the LIST command from the Inquiry toolbar. Select the argus logo in the lower right portion of the drawing. Press ENTER to complete the selection process. The AutoCAD Text window is displayed containing the objects data, as follows:

```
BLOCK REFERENCE    Layer: 1
Space: Model space
Color: 1 (red)     Linetype: CONTINUOUS
Handle = 22E1


LOGO
at point, X=101'-6 11/16"  Y=9'-5 1/4"  Z=    0'-0"
X scale factor    1.0000
Y scale factor    1.0000
rotation angle    0
Z scale factor    1.0000
```

6. This information shows that the logo and title block are a *BLOCK REFERENCE*. Continue to list other items in the drawing.

7. Restore the Office view (see the following figure) by using the Named Views button from the Standard toolbar.

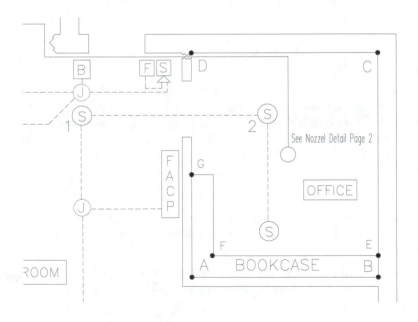

Figure 15-7: Office view of swb.dwg

8.   You are now going to use the DIST command to determine the linear dimensions for the office.

From the Inquiry toolbar, choose the Distance button. Select the endpoint of line AB, near A and then the endpoint of line CD, near C. The following values should be displayed at the Command prompt:

```
Distance = 15'-7 7/16",   Angle in XY Plane = 50,   Angle from XY
Plane = 0
Delta X = 10'-0",   Delta Y = 12'-0",   Delta Z = 0'-0"
```

The Distance is the absolute distance between points A and C. Delta X defines the length of the office and Delta Y defines the width. Therefore, the office is 10' x 12'.

9.   The DIST command can also be used to find the length of sprinkler pipe from the center of symbol S1 to the center of symbol S2.

From the Inquiry toolbar, choose the Distance button. Select the center of symbol S1 and then the center of symbol S2. The following values should be displayed at the Command prompt.

```
Distance = 9'-11 1/4",   Angle in XY Plane = 1,   Angle from XY
Plane = 0
Delta X = 9'-11 1/4",   Delta Y = 0'-1 1/16",   Delta Z = 0'-0
```

Notice that the symbols are not horizontal to the X axis. Delta Y verifies this.

10.   From the Inquiry toolbar, choose the Area button. You will obtain the area of the office by selecting the 4 corner locations ABCD in Figure 10-7. Select the following intersections:

▶   A

▶   B

▶   C

▶   D

Press ENTER to return the following values at the Command prompt.

```
Area = 17280.00 square in. (120.0000 square ft.), Perimeter =
44'-0"
```

The AREA command returned the values in square inches and feet, along with the perimeter of the room.

11.   This time you will obtain the floor area of the office with the bookcase removed. Because the area under the bookcase is not carpeted, the value returned will be the total area of carpet needed for the office.

From the Inquiry toolbar, choose the Area button. At the Command prompt enter **a** and press ENTER to activate Add mode.

12.   Select the following intersections in order:

▸ A

▸ B

▸ C

▸ D

Presss ENTER to return the following values and continue the command:

Area = 17280.00 square in. (120.0000 square ft.), Perimeter =
44'-0"

Total area = 17280.00 square in. (120.0000 square ft.)

<First point>/Object/Subtract:

13. At the Command prompt, enter **s** and press ENTER to activate Subtract mode. Then select the following intersections in order:

▸ A

▸ B

▸ E

▸ F

▸ G

▸ H

Press ENTER to return the following values and continue the command:

Area = 2383.25 square in. (16.5503 square ft.), Perimeter = 30'-
11 1/4"

Total area = 14896.75 square in. (103.4497 square ft.)

<First point>/Object/Add:

Press ENTER to complete the command.

14. From the Tools menu, choose Inquiry, then choose Time. The total editing time and elapsed time for the current drawing session is displayed in the AutoCAD Text Window.

15. From the Tools menu, choose Inquiry and then choose Set Variable. The SETVAR command is activated. Enter **?** for the Variable name and press ENTER twice to list <*> all system variables.

> Note: Exercise caution when changing system variable settings. Look up the system variable and become familiar with its purpose and settings before making any changes.

Conclusion

After completing this chapter, you have learned the following:

◗ You can display drawing statistics, modes, and extents in the command history area of the AutoCAD Text Window.

◗ The command that is used to locate the XYZ coordinates of a point location on the drawing is the id command.

◗ You can use the list command to List information about a specific object and display the data in the command history area of the AutoCAD Text Window.

◗ The distance command is to define the distance, angle, and deltas between to points.

◗ Use the area command to calculate the area and perimeter of objects or the combined area of more than one object using add and subtract modes.

◗ The time command is used to check for; time of original creation and latest revision, total editing time, and elapsed time for the current drawing session if the timer is reset.

◗ Listing and controling system variables can be done using the SETVAR command. You can also change system variable settings by entering the name of the system variable and its new value at the Command prompt.

Chapter 16

Applying Hatch Patterns

In this chapter, you learn how the AutoCAD BHATCH command greatly speeds up the task of employing hatching in drawings.

About This Chapter

In this chapter, you will do the following:

▶ Create a series of hatches with different patterns, scales and display properties.

▶ Specify new hatch patterns using existing patterns and properties as the basis for the new pattern.

▶ Edit associative hatch patterns, changing their boundaries, patterns and properties.

▶ Control hatch pattern visibility.

Hatch Patterns

Hatch patterns are used to fill enclosed areas in drawings for many purposes. They can be used to highlight or differentiate features, such as showing different materials in assemblies, rock types in maps, wall materials, foundations, gratings and other textured features. The following figure is an example of a simple drawing using two hatch patterns:

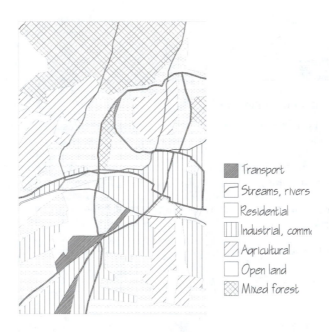

Figure 16-1: Example of drawing with hatch patterns

You can fill closed areas of nearly all types of AutoCAD geometry with *associative* hatch patterns, linked with the original defining boundary geometry. Associativity means that the relation between the hatch and the defining boundaries is stored, so that when you edit boundaries the pattern updates automatically. *Islands* are objects within outermost hatch boundaries that are hatched or not hatched as specified by the hatch style. You can also select an existing associative hatch pattern and change its pattern and properties without having to create a new hatch object. Associative hatching and internal boundary detection enhances your productivity by

> ▶ Letting you create complex hatch patterns quickly

> ▶ Providing the flexibility to change pattern designs by selecting from a range of

predefined hatch patterns

◗   Automating hatch pattern regeneration when the boundary changes

The Associativity and Island Detection options can be disabled to provide custom hatch patterns and properties.

Creating Hatch Objects

Three commands detect the internal boundaries in a closed region: BHATCH, BPOLY, and BOUNDARY. The BHATCH command fills the closed region with a hatch pattern; the pattern is a separate object.

With island detection, AutoCAD finds an outer boundary and checks to see if the "internal point" you selected is inside this boundary. If so, it selects all the geometry inside the outer border and determines which objects form a closed perimeter. The objects that do are highlighted to show which areas will be hatched.

When island detection is turned off, AutoCAD can only find a single boundary, which is then traced until a closed region is formed. In either case, once all the boundaries have been traced, they are converted to closed polylines or regions which are used as hatch borders.

Using the Boundary Hatch Dialog Box

Methods for opening the Boundary Hatch dialog box include:

◗   **Toolbar:** Draw

◗   **Menu:** Draw > Hatch

◗   **Command:** BHATCH

The Boundary Hatch dialog box is shown in the following figure:

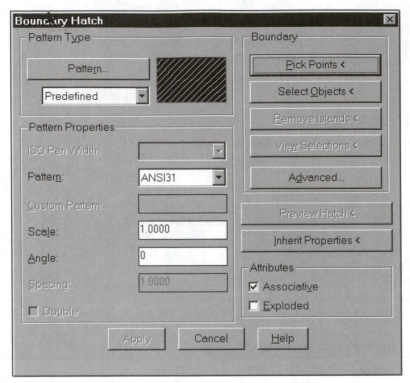

Figure 16-2: The Boundary Hatch dialog box

The three main steps in defining a hatch pattern are the following:

- ▶ Choosing a pattern to fill the area

- ▶ Defining the area to be hatched

- ▶ Specifying the properties of the hatch pattern

The options to control hatch pattern definition are provided in the Boundary Hatch dialog box.

Pattern Type

The Pattern Type area of the dialog box lets you choose the pattern to use. You can elect to use a supplied pattern by choosing Predefined. These patterns are stored in the standard *acad.pat* file. You can also choose Pattern to see a complete list of all the available patterns in the Hatch Pattern Palette dialog box. Initially, the hatch pattern is set to the last pattern used. The two other options for specifying a pattern are:

- ▶ User Defined sets the pattern to parallel lines with Spacing and Angle values.

- ▶ Custom lets you choose from a custom definition file.

Choosing Boundaries and Hatching Areas

The boundary area lets you define the area to be hatched. The simplest way to choose an area is with the Pick Points option. This temporarily dismisses the dialog box and prompts you for an internal point to find boundaries. If the area you choose is not a closed area, AutoCAD warns you and lets you try again. Once you have selected the area(s) to be hatched, press ENTER.

You can check the areas you choose for hatching with the View selection option.

> Note: Due to the nature of hatch associativity, it is recommended that you hatch only one area at a time. If you choose more, then all the areas are treated as one hatch pattern and you cannot edit them separately at a later time.

The other options in the Boundary area are:

- ▶ *Select Objects* - Lets you choose objects as boundaries

- ▶ *Remove Islands* - Lets you remove islands chosen in previous Pick Point or Select Object operations

Specifying Pattern Properties

This area lets you specify the way the hatch pattern that you selected in the Pattern Type area is displayed.

- ▶ *Pattern* - Sets the hatch pattern by name. If you choose an item from the Pattern Type area, the name of that pattern opens here.

- ▶ *Angle* - Sets the angle of the hatch pattern.

- ▶ *Scale* - Sets the scaling of the hatch pattern. Smaller values mean that the elements of the pattern are closer together.

The User defined pattern type lets you define a series of parallel lines using the current linetype as a hatch pattern. You can specify the separation of the lines with the Spacing option and the angle of the lines. You can also use the Double option to repeat the lines at a 90 degree angle.

Other Hatch Options

The following list describes other hatch options:

- ▶ *Associative* - When checked, the new hatch pattern is associated with its defining borders.

- ▶ *Advanced* - Shows the Advanced Options dialog box. This controls features such as object type, island detection, and boundary generation.

- ▶ *Exploded* - When checked, explodes the hatch pattern as you apply it.

Applying the Hatch

You can preview the hatch pattern by choosing the Preview Hatch button. This option temporarily dismisses the dialog box and draws the current hatch pattern, but does not add it to the database of your drawing. This is recommended because patterns may not be displayed as you expect, or may have an inappropriate scale, and drawing a hatch you do not want can take a long time. When the preview is complete, choose Continue.

Once you have defined all the required values, the Apply button becomes available. If you have seen that the hatch pattern is what you desire, choose Apply.

The hatch pattern you create is a new and separate object. The object stores the boundary of the hatch pattern with pointers. The pointers are linked to the geometry in the drawing database defining the pattern for each hatch object. The pattern is drawn with the current properties for color and linetype.

Each element of the hatch pattern is a separate object, so you can use object snaps, such as Midpoint or Endpoint, to snap to parts of the hatch pattern.

The Solid Fill Pattern

A solid fill completely fills an enclosed area so that the area looks like a solid color (a raster fill) rather than a series of closely-spaced lines. Solid fills are used extensively in mapping to indicate soil types, geology, districts and regions.   Solid fills can be used to demonstrate a closer representation of a finished product's color and texture.

The solid fill hatch pattern is created just like any other hatch pattern, except that:

> ◗ The solid pattern in the Pattern area displays the current color settings.

> ◗ All other Pattern properties are disabled.

> ◗ Solid fill is displayed even if the current linetype is not continuous.

You can use the Pick Points or Select Objects options in the Boundary Hatch dialog box to indicate solid fill boundaries. When using the Select Objects option the boundaries must be clearly defined.

Note: The following system variables are related to the BHATCH command:

> ◗ HPANG - The current hatch angle

> ◗ HPBOUND - Type of boundary created; 0 = Polyline, 1 = Region

> ◗ HPDOUBLE - User-defined double lines

> ◗ HPNAME - The default hatch pattern

> ◗ HPSCALE - Default scaling for user-defined patterns

> ◗ HPSPACE - Default line spacing for user-defined patterns

Exercise 16-1: Creating Boundary Hatches

In this exercise, you use AutoCAD's associative hatching capabilities to add hatch patterns to complete an existing drawing.

Creating Boundary Hatches

1. Open the file *plane.dwg*. The drawing looks like the following figure:

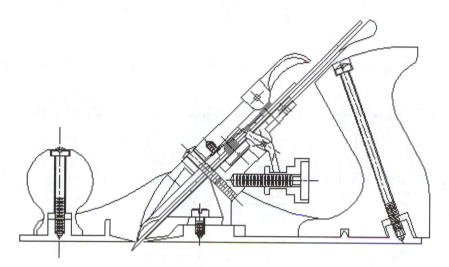

Figure 16-3: Jack plane assembly

The drawing file is an assembly drawing showing the half-section of a jack plane. Using just the BHATCH command, you can create all the necessary hatch patterns to create a final drawing.

Each major component of the jack plane assembly is organized on a layer whose name reflects the name of the component.

If necessary, you can change your layer visibility settings for each component to make hatching the plane easier.

2. Freeze the layers FASTENERS and CENTER. Set the current layer to HATCH-PATTERNS. This is the layer on which you produce the hatched details.

3. Start the BHATCH command by choosing Hatch from the Draw menu.

The Boundary Hatch dialog box is displayed.

4. Check that the Pattern Type is set to Predefined.

In the Pattern Properties area, check that the style is set to Ansi31 with a scale of 1.0 and an angle of 0.

To apply hatching to the drawing, choose the Pick Points button, then select the points P1 and P2 on the knob, as shown in the following figure:

297

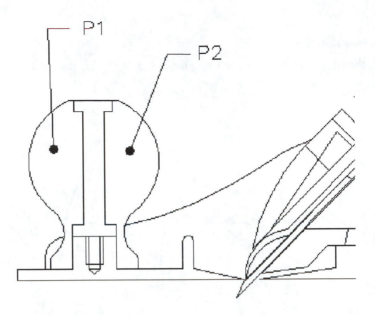

Figure 16-4: Two internal points for the knob

5.   Press ENTER to complete selection and return to the dialog box.

In the dialog box, choose the Preview Hatch button to preview the hatching.

Choose the Continue button to return to the dialog box.

At this stage, the hatch pattern has not been added to the drawing and you can still change the specified pattern, scale, and angle.

6.   To add the hatching to the knob, choose Apply.

The result is shown in the following figure:

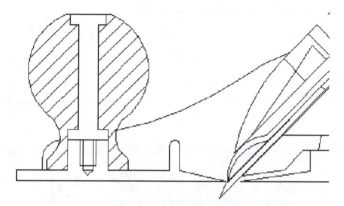

Figure 16-5: Hatching applied to the knob

7. Press ENTER to repeat the command.

In the Pattern Properties area, select the Plasti Pattern by scrolling down the list. Enter a scale of **2.0** and an angle of **45** degrees.

To define the areas to be hatched on the handle of the plane, choose the Pick Points button, choose point P1, and press ENTER to return to the dialog box. Preview the hatch, then apply it, as shown in the following figure:

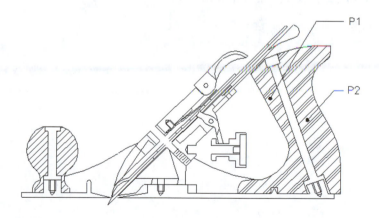

Figure 16-6: The hatched handle

8. Repeat the BHATCH command and choose the Select Objects button. Select the line

surrounding point P2, as shown in the previous figure. Press ENTER to return to the dialog box. Preview the hatch, then apply it.

9.   Repeat the BHATCH command and hatch the base of the plane.

Use the Pattern option to specify the Ansi32 pattern. Choose OK.

Enter a Scale of **0.75** and an Angle of **75** degrees.

Select the pick points P1 and P2 shown in the following figure to define the areas to hatch.

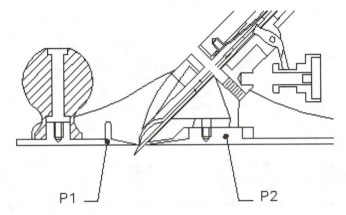

P1   P2

Figure 16-7: Pick points for hatching the base of the plane

10.   Preview the hatch, then apply it.

The result should look like the following figure:

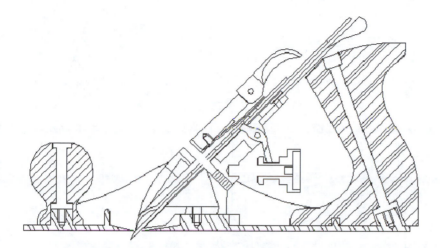

Figure 16-8: The hatched base of the plane

As well as being able to define a hatched area by simply selecting an area and 'flood filling' to its boundaries, AutoCAD also provides the option to hatch selected objects. Now, you hatch the thumb knob on the plane. The two respective regions which represent the cross section of this component have been drawn as closed areas.

Hatching the Thumb Knob

1. Zoom in on the area shown in the following figure:

2. Start the HATCH command, and specify the hatch pattern as Ansi31 with a scale of **0.5** and an angle of **120** degrees.

    Instead of using Pick Points, choose the Select Objects button and select the two polyline outlines, as shown at P1 and P2 in the following figure:

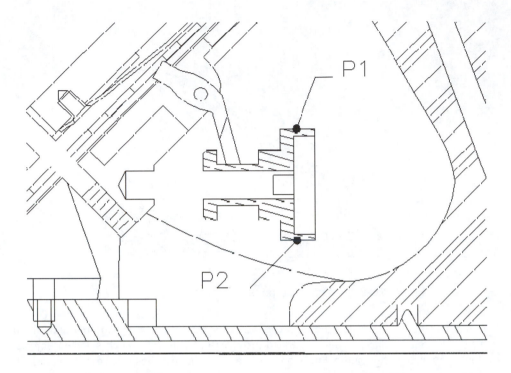

Figure 16-9: The two areas to be hatched

3.   Preview the hatch and then apply it.

4.   Adjust the view of the plane so that the 'mount plate' of the plane can be seen.

5.   Repeat the BHATCH command, and using the same hatch pattern, scale and angle, hatch the mount plate using the Pick Points option in the areas shown in the following figure:

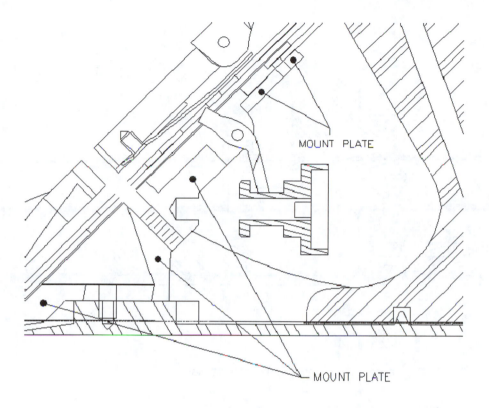

MOUNT PLATE

MOUNT PLATE

Figure 16-10: Areas to be hatched

6.   The completed hatching looks like the following figure:

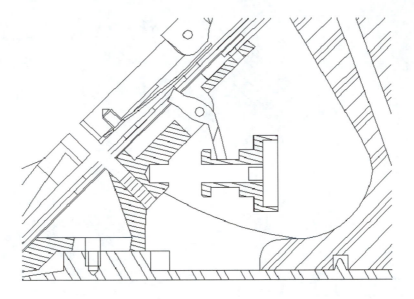

Figure 16-11: The hatched mount plate

7.　　Save your drawing.

Chapter Summary

After completing this chapter, you have learned the following:

▶   The BHATCH command fills the closed region with a hatch pattern.

▶   Editing an associative hatch pattern means that the relationship between the pattern and its boundaries is retained when the boundaries are changed.

▶   Existing patterns can be edited or used as the basis for new patterns.

▶   You can use grips to modify a hatch pattern and its boundaries.

▶   Hatch pattern visibility can be turned on and off with the FILL command.

Chapter 17

Editing Hatch Patterns

In this chapter, you learn how to edit hatch patterns, properties, boundaries, and how to use these as the basis for other hatch patterns after their creation.

About This Chapter

In this chapter, you will do the following:

▶ Edit associative hatch patterns, changing their boundaries, patterns and properties.

▶ Control hatch pattern visibility.

Associative Hatch Patterns

You can fill closed regions of nearly all types of AutoCAD geometry with *associative* hatch patterns. Associativity means that the relation between the hatch and the defining boundaries are stored, so that when you change boundaries the pattern updates automatically.

You can also select an existing Associative Hatch pattern and change its pattern type and properties without having to create a new hatch object.

Using the Hatchedit Dialog Box

The dialog box interface for the HATCHEDIT command is the same as the BHATCH command although some options remain greyed out. Solid fill uses the same dialog box for editing.

Methods for invoking the HATCHEDIT command include:

- ▶ **Toolbar:** Modify II
- ▶ **Menu:** Modify > Objects > Hatch
- ▶ **Command:** HATCHEDIT

Then choose an existing hatch pattern using the Hatchedit dialog box, which is shown in the following figure:

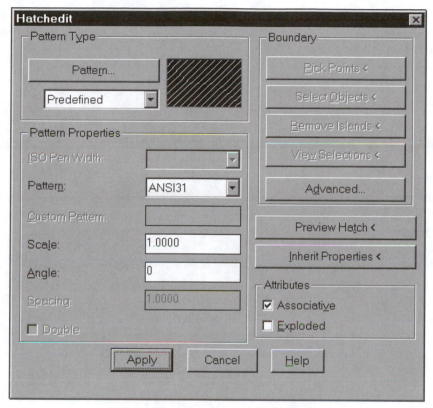

Figure 17-1: The Hatchedit dialog box

Features of Editing Associative Hatch Patterns

The following features of editing hatch patterns and boundaries are a result of creating associative hatch patterns. Any Modify command that alters the hatch boundary causes the hatch object to regenerate if the final boundaries remain closed. For example, you can use the PEDIT command to curve fit a previously-hatched polyline border. However, some Edit command operations, especially STRETCH, can result in open boundaries when segments of the boundary are removed from the selection set or deleted as part of the edit operation. An associative hatch pattern is updated in a block reference when the defining block boundary is redefined.

If you want to erase an associative hatch pattern and not the boundary, make sure PICKSTYLE is set to 0 or 1. If PICKSTYLE is 2 or 3, erasing the pattern also erases the boundary. When you use the EXPLODE command to explode an associative hatch object, it removes the association between the hatch pattern and its defining boundary and explodes the hatch block into linear elements. You cannot explode a solid fill. Associativity is preserved when an associative hatch object is copied, moved, or mirrored, provided all its boundary segments are selected with it. If any part of the boundary is missing, associativity is removed. However, if the hatch pattern is on a layer that is locked or frozen, the hatch pattern is not updated if the boundaries are modified.

You can select and edit individual boundaries, such as islands, in a hatched region. Hatch patterns remain associative if you delete part of the boundary, and the remaining boundary forms a closed area, as shown in the following figure:

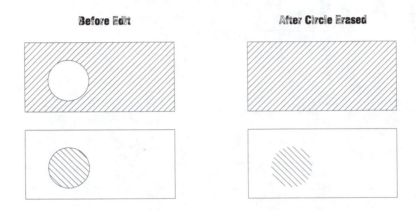

Figure 17-2: Edit hatch pattern boundaries

If you select the pattern and part of the boundary with the MOVE, SCALE, STRETCH, and ROTATE commands, the boundary objects are edited, and the hatch is recalculated. If the boundary is still valid, the hatch is updated and remains associated with the boundary. If the boundary is not valid, the hatch is disassociated, and the edit applies to the pattern only. An example of such an edit is shown in the following figure. The top row displays examples of editing an island in the pattern (text). The bottom row shows the effects of selections of the boundary and the pattern.

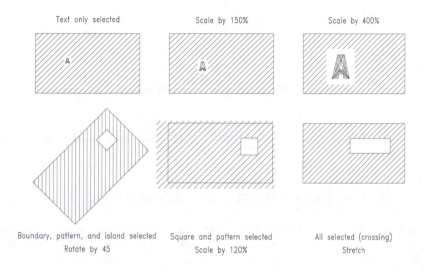

Figure 17-3: Edit hatch patterns, boundaries and islands

If you select the pattern and part of the boundary with the COPY, ARRAY, and MIRROR commands, you create a new non-associated hatch object.

You can use the HATCHEDIT and DDMODIFY commands to edit both associative and non-associative patterns. You can also Preview Hatch from within the HATCHEDIT dialog box. If you use DDMODIFY to modify a hatch pattern or its boundary, the boundary is recalculated and the pattern remains associative if the boundary is valid.

You cannot use these commands to edit hatch objects: FILLET, CHAMFER, OFFSET, BREAK, MEASURE, and DIVIDE. You can change the layer, color, linetype and linetype scale of a hatch pattern with the DDMODIFY and DDCHPROP commands.

Editing Hatch Patterns with Grips

You can modify associative hatch patterns with object grips when object grips are enabled. Use the DDGRIPS command to enable object grips. If you want to directly edit the hatch pattern, make sure the Enable Grips Within Blocks option is selected.

If you select an associative hatch pattern, it is highlighted and shows a grip at the centriod of the hatch pattern. The centroid is at the center of the bounding extents of the hatch patterns that were created at one time. The grip can also be displayed outside the hatched area. Examples of hatch pattern centroids are shown in the following figure:

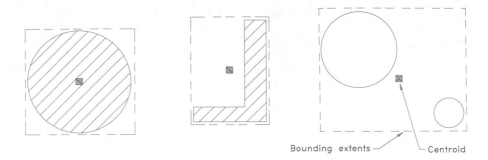

Figure 17-4: Examples of centroids for hatch patterns

If you select the boundary of a hatch pattern, the pattern is highlighted, and grips are displayed at each vertex of the boundary. When you finish a grip edit operation, the hatch pattern adapts to the changes if your editing left a closed boundary. If the resulting boundary is *not* closed, the hatch pattern is *not* modified and associativity is removed from the hatch pattern. In these cases AutoCAD disassociates the hatch pattern with its defining geometry and a warning message is displayed:

hatch boundary associativity removed

Hatch Visibility

Pattern visibility can be controlled by entering **fill** at the Command prompt, then entering **on** or **off**. The system variable FILLMODE can also be used to control pattern visibility. To make all patterns invisible, enter **fillmode** at the Command prompt and set it to **0**. You must perform a regeneration (REGEN command) before the change takes place.

You can edit pattern boundaries when the patterns are not visible. AutoCAD shows the message "Analyzing associative hatch..." and maintains changes made to pattern boundaries. If you break a closed pattern area, you see the message "Hatch boundary associativity removed." You can view the modified patterns by setting FILLMODE to 1 and carrying out a regeneration.

FILLMODE also controls the visibility of objects created with the SOLID command and the filled display of wide polylines. You can control the visibility of hatch objects by freezing or turning off layers. You can also lock layers with hatch patterns. However, you must turn on, thaw, or unlock hatch layers in order for associative hatch patterns to reflect pattern boundary changes. Pattern visibility can also be controlled by using FILL instead of FILLMODE and selecting the On or Off option.

Converting Hatch Patterns

When you open a pre-Release 14 drawing that contains hatch patterns, AutoCAD does not automatically convert the patterns to Release 14 associative hatch objects, although the HATCHEDIT command does convert a selected Release 13 hatch pattern.

With the CONVERT command, you can manually convert any pre-Release 14 two-dimensional polylines or hatch patterns to Release 14 optimized polylines and hatch objects respectively.

Exercise 17-1: Modifying Boundary Hatches

 In this exercise, you look at using an existing boundary hatch as the basis for defining a new hatch pattern. This is termed inheriting properties. You also look at editing associative hatch patterns.

Using Inherited Properties of Hatch Patterns

AutoCAD lets you choose a hatch pattern definition by selecting an existing pattern in the drawing. This saves you from having to remember the respective hatch name, scale and angle if you want to use it again.

1. Adjust the view of the plane so that the wedge iron is fully in view, as shown in the following figure:

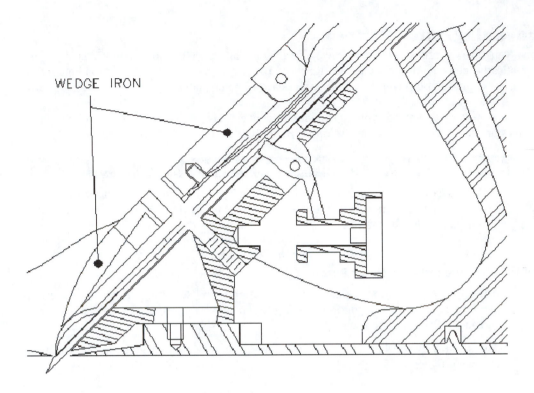

WEDGE IRON

Figure 17-5: The wedge iron

2.   Start the BHATCH command and choose Inherit Properties.

In the drawing you are prompted to select a hatch pattern.

3.   Select the hatch pattern you used for the base of the plan.

In the dialog box, the hatch pattern definition is updated.

Using the Pick Points option, select the two areas of the wedge iron shown in the previous figure, and apply the hatching.

The result is shown in the following figure:

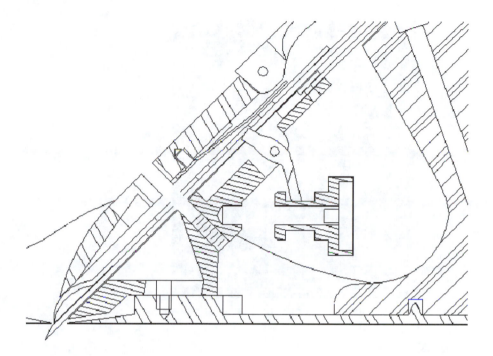

Figure 17-6: The hatched wedge iron

Hatch Patterns and Associativity

You can fill nearly all types of AutoCAD geometry with associative hatch patterns. This means that the relationship between the hatch and its defining boundaries is stored. If the boundaries are changed, the hatching is automatically updated. You are now going to change the geometry of the handle to demonstrate how the associative hatching is updated.

1.   From the Tools menu choose Grips.

In the Grips dialog box, make sure that the Enable Grips checkbox is checked.

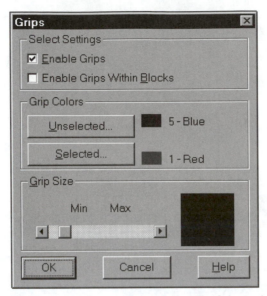

Figure 17-7: Grips dialog box with Enable Grips selected

2. Choose OK.

3. Change your view so you can see the handle of the plane.

Select the profile on the right hand side of the handle, taking care not to select the hatch. The grips on the profile are displayed and the profile becomes highlighted.

4. Make one of the grips on the handle profile hot and then stretch it to redesign the profile.
> Note that the hatching is automatically updated each time you move the grips.

5. Press ESC twice to end the editing and remove the grips.

Editing Hatch Patterns

1. Start the HATCHEDIT command by choosing Objects from the Modify menu. Then choose Hatch.

At the prompt for `Select hatch object`, select the hatch pattern you just edited with grips.

2. Now, that the profile has been redesigned, enter the scale as **1.5** and an angle of **75** degrees. Then Apply the hatch.

The changed hatch pattern is shown in the following figure:

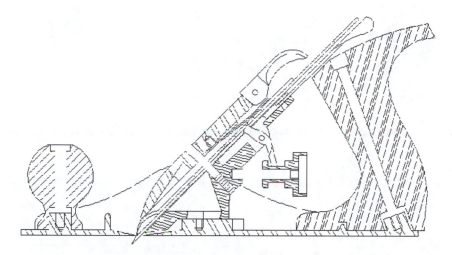

Figure 17-8: Edited hatch scale on handle

Controlling Hatch Pattern Visibility

1. Start the FILL command by entering **fill** at the Command prompt. Then enter **off** to hide all hatch patterns in the drawing.

    To show the changed state, you need to regenerate the drawing.

2. From the View menu, choose Regen.

    All hatch patterns in the drawing are now hidden from view.

3. To restore the hatch pattern visibility, enter **fill** and set it to **on**. Then regenerate the drawing.

4. Save your drawing.

Conclusion

After completing this chapter, you have learned the following:

> ▶ The BHATCH command fills the closed region with a hatch pattern.

> ▶ Editing an associative hatch pattern means that the relationship between the pattern and its boundaries is retained when the boundaries are changed.

> ▶ Existing patterns can be edited or used as the basis for new patterns.

> ▶ You can use grips to modify a hatch pattern and its boundaries.

> ▶ Hatch pattern visibility can be turned on and off with the FILL command.

Chapter 18

Annotating a Drawing

Annotating a drawing lets you present information that cannot be presented graphically. This information can be included as text in dimensions, notes, and titles. AutoCAD provides you with a number of techniques that let you effectively create and place text in a drawing.

About This Chapter

In this chapter, you will do the following:

▶ Create new text styles using STYLE.

▶ Place text in drawing using DTEXT.

Creating New Text Styles

Placing text in a drawing requires that you use a *text style*. AutoCAD provides one style, STANDARD, which is the default text style in the *acad.dwt* and *acadiso.dwt* template files. If STANDARD is not suitable for the needs of your project, then you can create other text styles. Text styles are easily created using the STYLE command.

Creating Text Styles Using STYLE

A text style in AutoCAD sets the font, size, angle, and other characteristics as shown in the following table:

Setting	Default	Description
Style name	STANDARD	Name of style, up to 31 characters
Font name	*txt.shx*	Name of specified font
Font style	None	Font formatting
Height	0	Text height
Use Big Font	No	Special shape definitions
Width factor	1	Condensed < 1 or expanded > 1
Oblique angle	0	Slants the text between -85° and 85°
Backwards	No	Text is placed backwards
Upside-down	No	Text is placed upside-down
Vertical	No	Text is placed vertically

Table 18-1: Settings for STANDARD text style

In most applications the STANDARD text style described in Table 18-1 will be modified. To create new text styles or modify an existing style you will use the STYLE command.

▶   **Menu:** Format > Text Style

▶   **Command:** STYLE

The STYLE command will display the Text Style dialog box, as shown in the following figure:

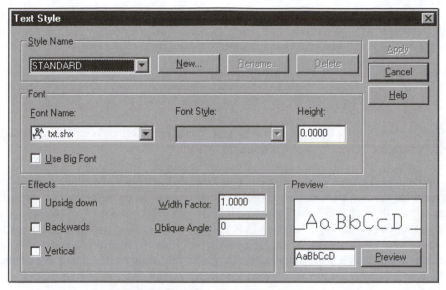

Figure 18-1: The Text Style dialog box

The Style Name area displays the name of the current text style in a drop-down list. You can create a New style, Rename a current style, or Delete a current style using the respective buttons.

Note: The Rename and Delete options are not available for the STANDARD style.

Selecting New displays the New Text Style dialog box, as shown in the following figure:

Figure 18-2: The New Text Style dialog box

You can select a Font Name from the Font Name drop-down list, as shown in the following figure:

Figure 18-3: The Font Name drop-down list

The font names listed are the registered TrueType fonts and the AutoCAD compiled SHX fonts which are in the AutoCAD \fonts directory. Each font name is preceded by an icon indicating whether it is a TrueType or AutoCAD font, as shown in the following figure:

Figure 18-4: The TrueType and AutoCAD icons

TrueType fonts let you work with text in a way that is similar to other Windows applications. They allow for character formatting such as **bold,** *italic,* and <u>underscore</u>.

The Height option lets you set a height for the new or modified text style. Setting a defined height for each text style is another method of adhering to standards. If the drawing you are working on requires annotations that are 2, 3, and 5 units high, then you can create new text styles for each height and give the style a meaningful name. They could be named *gen_text2* for general text that is 2 units high, *t_block3* for entries in the title-block which are 3 units high, and *t_block5* for entries in the title-block which are 5 units high. The list of text styles will expand as you require new styles for dimensioning, general notes, bill of materials, schedules, and specifications.

If you create text styles with 0 height, AutoCAD will prompt you for the height each time you create line text. If you want to set the text height as you place the text then you should create the style with a height of 0. The Use Big Font option is required if you are using an Asian-language Big Font file. This option can only be used with AutoCAD's SHX fonts.

The options in the Effects area let you modify characteristics of the font. You can specify whether or not you want a font displayed Upside Down, Backwards, or Vertical, as shown in the following figure:

Figure 18-5: Text placed upside down, backwards, and vertical

The Width Factor expands or compresses each character. A value < 1 compresses the character while a value > 1 expands the character as shown in the following figure:

Figure 18-6: Text placed width factor .5 and 1.5

The Oblique Angle lets you set an angle between -85 and 85 degrees measured from the vertical. This option can be used in conjunction with the ROTATION command to create text styles for annotation on an isometric layout, as shown in the following figure:

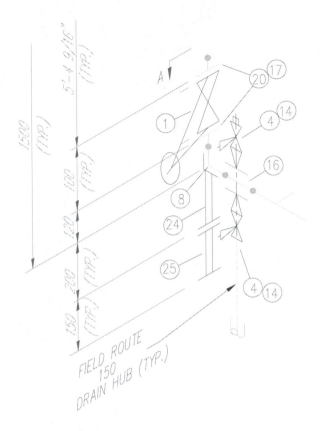

Figure 18-7: Rotated text with an oblique value

The Preview area displays sample text. It defaults to a series of letters as shown in Figure 12-1, but you can enter different sample text, as shown in the following figure:

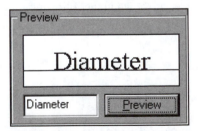

Figure 18-8: Changing the sample text

When you open the Text Style dialog box the Apply button is not available for selection. The Cancel and Help buttons are available. Choosing Cancel before you make any changes to the current style will exit the dialog box. Any changes to the options in the Style name area will change Cancel to Close. Changes in the Font or effects areas will make the Apply button available for selection. Choosing Apply will cause the changes to the current style to take effect in the drawing.

Placing Text in a Drawing

Text in a drawing can generally be divided into two categories: single lines of text such as the annotation describing an orthographic view, or paragraph text such as general notes. AutoCAD provides the DTEXT command for the placement of single lines of text, and the MTEXT command for paragraph text.

DTEXT Command

DTEXT or Dynamic Text places single line of text in a drawing as you enter the text string at the Command prompt. You can also place several lines of text without exiting the command.

Methods for invoking the DTEXT command include:

▶ **Menu:** Draw > Text > Single Line

▶ **Command:** DTEXT

The DTEXT command initially prompts you for text justification, style, and the start point for the line of text, as follows:

```
Command: _dtext Justify/Style/<Start point>:
Height <2.5000>:
Rotation angle <0>:
Text: Section A-A
Text:
```

Justification

The Justify option controls the alignment of the text. Selecting Justify will display the following options for controlling the placement of the text:

```
Command: dtext
Justify/Style/<Start point>: j
Align/Fit/Center/Middle/Right/TL/TC/TR/ML/MC/MR/BL/BC/BR:
```

▶ *Align* - Controls the height and orientation of the text based on the two points you select. The text height will become shorter as the text string becomes longer.

▶ *Fit* - Places text in the drawing based on the two points you select and the specified height. Unlike the Align option, the text height remains constant while the width of the characters adjusts depending on the length of the text string. Fit will only place text in a horizontal alignment.

▶ *Center* - Places text in the drawing based on the point you specify, the height, and rotation, as shown in the following figure

▶ *Middle* - Places text based on the point you specify, the height, and rotation as shown in the following figure

▶ *Right* - Places text based on the point you specify, the height, and rotation, as shown in the following figure

If you specify a point on the screen for rotation, then the text baseline is aligned through the center point and the rotation point. A rotation point placed to the left of the center point will place the text upside-down.

As well as the text placement options discussed, you can specify justification based upon the 9 locations shown in the following figure:

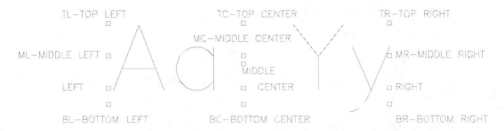

Figure 18-9: Text justification options

Note: The default justification is left.

If you create a text style with a height of 0 the Height prompt will be displayed each time you place line text. The Rotation prompt lets you place the text at a specified angle.

Creating Special Characters

Placing text in an AutoCAD drawing can be enhanced with the addition of special characters such as the degree symbol, plus/minus tolerance symbol, and the diameter symbol. You can place these characters using either *Unicode Characters* or *Control Codes*.

Unicode character strings are shown in the following table:

Unicode String	Purpose	Input Example	Output
\U+00B0	Degrees symbol	45.71\U+00B0	45.71°
\U+00B1	Tolerance symbol	45.71\U+00B1 .01	45.71±.01
\U+2205	Diameter symbol	\U+2205 45.71	Ø 45.71

Table 18-2 : Unicode characters

Further information about Unicode characters is available in Chapter 3 of the *Customization Guide* in the "Shapes, Fonts, and PostScript Support" section, and in Appendix A of the *User's*

Guide.

Control codes provide for 5 special characters, as shown in the following table:

Code	Purpose	Input Example	Output
%%o	Overscore On/Off	%%o45.71	45.71
%%u	Underscore On/Off	%%u45.71	<u>45.71</u>
%%d	Degrees symbol (°)	45.71%%d	45.71°
%%p	Tolerance symbol (±)	45.71 %%p .01	45.71 ± .01
%%c	Diameter symbol (Ø)	%%c 45.71	Ø 45.71

Table 18-3: Control codes

Placing Paragraph Text in a Drawing

Placing paragraph text in AutoCAD is done using the MTEXT command.

Multiline Text Editor

Options in accessing the MTEXT command include:

▶ **Toolbar:** Draw

▶ **Menu:** Draw > Text > Multiline Text

▶ **Command:** MTEXT

When you use the MTEXT command, you are prompted for the corners of a rectangle.

```
Command: _mtext
Current text style: STANDARD. Text height: 2.5
Specify first corner:
Specify opposite corner or[Height/Justify/Rotation/Style/Width]:
```

MTEXT uses the rectangle you define, fitting the text within the sides of the rectangle. Text can flow beyond the top and bottom of the rectangle, depending upon the justification selected. The insertion point of the text is based upon one of the nine justification points discussed previously. This insertion point is not necessarily one of the points that defined the rectangle. To help with the visualization of this process, arrows are displayed within the rectangle to indicate which direction the text will flow when the command is complete.

The following figure illustrates this process:

Figure 18-10: Mᴛᴇxᴛ text flow

When the opposite corner is selected, the Multiline Text Editor is displayed. The Multiline Text Editor dialog box, shown in the following figure, has three tabs. Depending upon the active tab, the controls in the horizontal display are different. However, the text editor area and the buttons on the right side of the dialog box are always displayed.

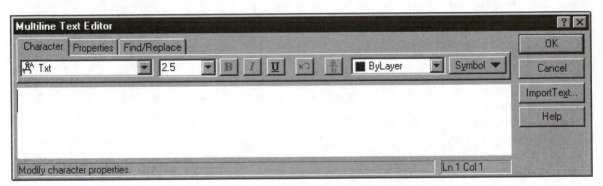

Figure 18-11: The Multiline Text Editor dialog box

The initial size of the dialog box is minimal, displaying space for approximately three lines of text. The dialog box will increase in size as more text is entered, up to three quarters of the screen height. A scroll bar is then displayed for larger blocks of text.

The text editor has a number of features. In general, these features reflect the techniques used in most word processors. Examples of this are the three ways that text can be highlighted: selecting the text by holding down the left mouse button and dragging, double-clicking to select the entire word, or triple-clicking to select the paragraph. The Character controls display the text formatting of the selected text, assuming that the formatting is consistent within the selection. As in most word processors, right-clicking the mouse displays a cursor menu, as shown in the following figure:

Figure 18-12: The right-click cursor menu

The Character Tab

The Character tab provides you with control of font, text height, bold, italic, underscore, stacked text, undo, color, and insertion of special characters, as shown in the following figure:

Figure 18-13: The Character Tab

The current font is displayed, as shown in Figure 18-13. A new font can be selected, or highlighted text can be changed by selecting a different font from the drop-down list, as shown in the following figure:

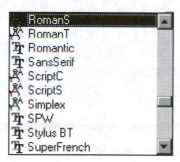

Figure 18-14: The Font drop-down list

The font height of the current text style is displayed. Since multiline text objects can contain text with different heights, all heights will be displayed in the Font Height drop-down list. Characters or text strings using TrueType fonts can be formatted using the Bold, Italic and Underline buttons. SHX fonts support underline only. The Undo button undoes the last edit in the Multiline Text Editor.

You can stack or unstack text that contains either the carat symbol (^) or a forward slash (/) using the Stack/Unstack option. If neither the / or ^ are present, then the stacking button is not highlighted. The stacked text is displayed at 70 percent of its height and is stacked evenly above and below each other, midway from the top and baseline of the text. Text to the left of the symbol is placed on top of the text. Typically, the carat is used for left-justified tolerance values, while the forward slash is used for center justified fractional numbers.

Text color can be set for new text or the color of existing text can be changed using the Text color option. The drop-down list contains options for Bylayer, Byblock, the first seven AutoCAD colors, the four most recently used colors, and an Other option which displays the Select Color dialog box.

The Symbol drop-down list contains 3 of the control code characters, a non-breaking space, and Other as shown in the following figure:

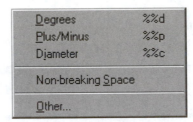

Figure 18-15: The Symbol drop-down list

The Other option displays the Unicode Character Map dialog box, as shown in the following figure:

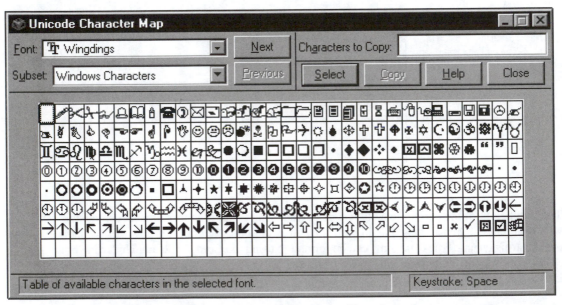

Figure 18-16: The Unicode Character Map dialog box

If you want to insert a character from the dialog box, select it, and then copy and paste it into the Multiline Text Editor dialog box.

The Properties Tab

You can access the properties of an MTEXT object from the Properties tab, as shown in the following figure. Properties apply to the whole object, unlike character formatting, which lets you edit portions of the object.

Figure 18-17: The Properties Tab

The STANDARD text style is the default style available in all drawings. You can create other styles during a drawing session, or provide the styles in template files. All available styles are displayed in the Style drop-down list, however certain characteristics may not be displayed in the text editor. Only SHX fonts support vertical text, and backward and upside-down styles are only available as Text objects. MTEXT ignores these characteristics. This is particularly important when you are applying changes to text using Properties, since this option affects the whole object. Character formatting applied to parts of the object is lost. The following table shows how specific overrides and properties are affected:

Overrides Affected	Overrides Not Affected	Properties Not Affected
Font	Underline	Alignment
Height	Color	Width
Bold Italic	Stacking	Rotation

Table 18-4: Affected overrides and properties

Text justification is easily selected from the Justification drop-down list as shown in the following figure:

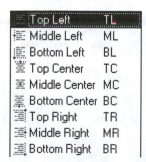

Figure 18-18: Selecting text justification

The width of the MTEXT object can be changed directly using the Width drop-down list as shown in figure 18-19. The values displayed are widths of MTEXT objects placed during the current drawing session. The No Wrap option is the same as a width of 0.

Figure 18-19: Setting the Mtext width

The Rotation drop-down list provides you with predefined angles and the option of directly entering a rotation value. The value shown reflects the current values set in Units.

The Find/Replace Tab

The usability of MTEXT is enhanced with the Find/Replace feature. This feature is similar in operation to existing find/replace capabilities found in word processors. These include Find, Find and Replace with Match Case, and Whole Word options, as shown in the following figure:

Figure 18-20: The Find/Replace Tab

Creating text objects in AutoCAD is easily accomplished. However, in many instances the desired text exists as a text file in another software application. The Import option of the MTEXT command lets you bring in an ASCII *.txt* file, or an *.rtf* file. Text formatting, such as bolding, will be retained if the inserted file is an *.rtf* file. This is also the case for text that is cut and paste from the Windows clipboard.

Selecting text prior to importing a text file causes the Inserting Text dialog box to be displayed, as shown in the following figure. If no text is selected, then the text is placed at the cursor location.

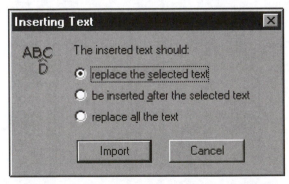

Figure 18-21: The Inserting Text dialog box

The file that is being imported cannot exceed 16K. An Alert box is displayed if the file size is too large. Also, a dialog box is displayed to inform you if the file contains non-text characters, and that the file insertion will be canceled. This is likely to occur when a file is saved in the native format, such as a Word *.doc* file.

Exercise 18-1: Defining and Using Text

Single line text can be placed in an AutoCAD drawing using the DTEXT command, and paragraph text can be inserted using the MTEXT command. In this exercise you will create new text styles using the STYLE command, and place text using DTEXT and MTEXT. You will also import an existing text file.

Creating New Text Styles

AutoCAD supplies one text style named STANDARD. In most drawings you will need to create new text styles for the different annotations that are required. In this section of the exercise you will use the STYLE command to create new styles for a drawing you have been asked to complete.

1. Open the file *cover.dwg*. The drawing looks like the following figure:

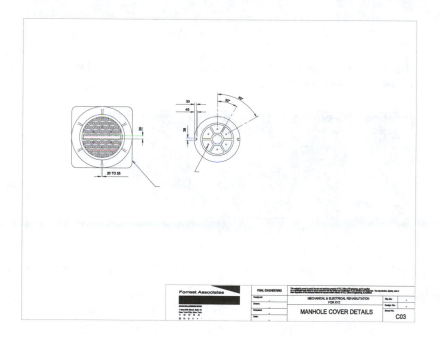

Figure 18-22: The incomplete cover.dwg

2. From the Format menu, choose Text Style. The Text Style dialog box is displayed, as shown in the following figure:

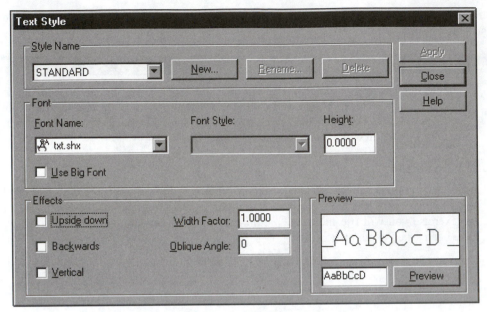

Figure 18-23: The Text Style dialog box

3. Select the New button in the Style Name section. The New Text Style dialog is displayed, as shown in the following figure:

Figure 18-24: The New Text Style dialog box

Enter **text5** in the Style Name field, then choose OK. The new style name is displayed.

4. Complete the settings for the new style as shown in the following figure:

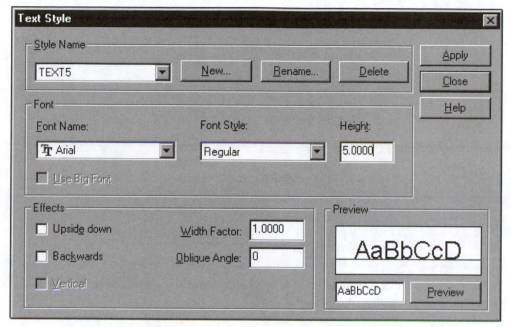

Figure 18-25: The settings for the new style

Then choose Apply. You have created a new text style named TEXT5, which is now the current style.

5. Repeat this process and create a new text style using the settings shown in the following table:

Name	Font Name	Font Style	Height
TEXT3	TT Arial	Regular	3

Table 18-5: New Text style settings

6. Then choose Close.

7. From the View menu, choose Named Views. Select V1, then select Restore. Then Choose OK. Your drawing should look like the following figure:

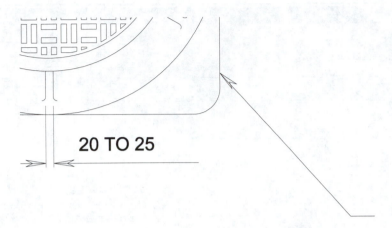

Figure 18-26: View V1

8.   From the Draw menu, choose Text. Then choose Single Line Text. Enter the following string of options at the Command prompt:

Command: _dtext Justify/Style/<Start point>: **j**

Align/Fit/Center/Middle/Right/TL/TC/TR/ML/MC/MR/BL/BC/BR: **ml**

Middle/left point: select the snap point to the right of the horizontal line of the leader

Rotation angle <0>:

Text: **USE SQUARE BASE**

Text: **OVER SQUARE**

Text: **OPENINGS**

Text: Press Enter to end placing text

Your drawing should now look like the following figure:

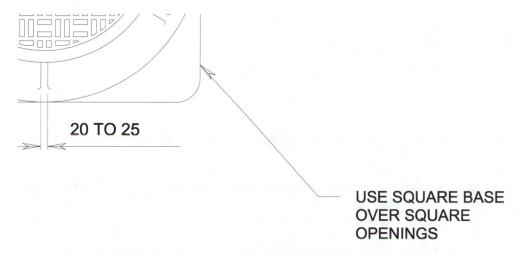

20 TO 25

USE SQUARE BASE
OVER SQUARE
OPENINGS

Figure 18-27: The completed text

Placing Centered Text

1. From the View menu, choose Named Views. Select V2, then choose Restore. Then Choose OK. Your drawing should look like the following figure:

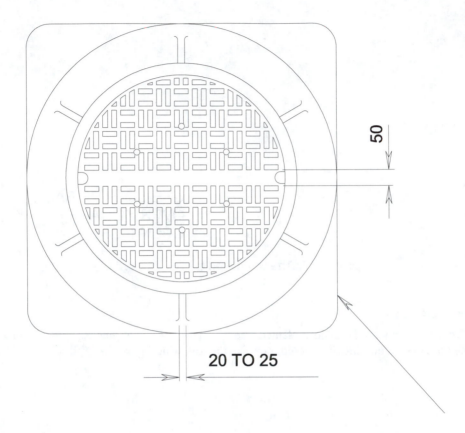

Figure 18-28: View V2

2. From the Draw menu, choose Text. Then choose Single Line Text. Enter the following string of options at the Command prompt:

Command: _dtext

Justify/Style/<Start point>: **j**

Align/Fit/Center/Middle/Right/TL/TC/TR/ML/MC/MR/BL/BC/BR: **mc**

Center point: _cen of select outside circle on the cover

Rotation angle <0>:

Text: **MANHOLE COVER**

Text: Press Enter to end placing text

Your drawing should look like the following figure:

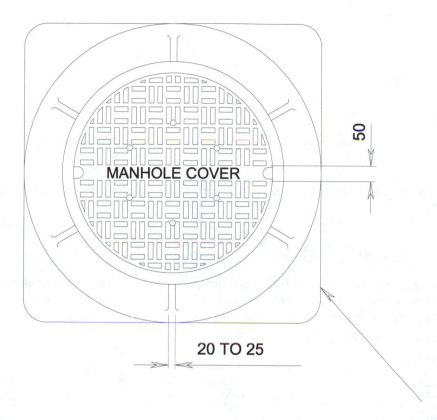

Figure 18-29: The centered text

Using Control Codes

1. From the View menu, choose Named Views. Select V3, then select Restore. Then Choose OK. Your drawing should look like the following figure:

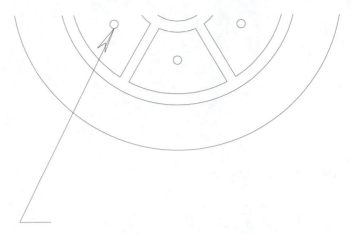

Figure 18-30: View V3

2. From the Draw menu, choose Text. Then choose Single Line Text. You will be placing text that contains a diameter symbol, so you will use the %%c code. Enter the following string of options at the Command prompt:

Command: **_dtext**

Justify/Style/<Start point>: **j**

Align/Fit/Center/Middle/Right/TL/TC/TR/ML/MC/MR/BL/BC/BR: **ml**

Middle/left point: select the snap point to the right of the horizontal line

Rotation angle <0>:

Text: **%%c20 HOLES (6)**

Text:

Your drawing should look like the following figure:

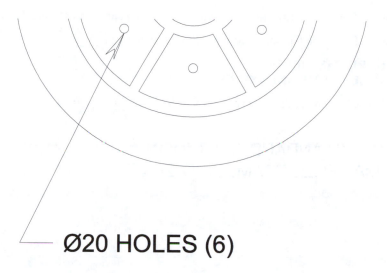

Ø20 HOLES (6)

Figure 18-31: The completed text

Placing Multiline Text

1. From the View menu, choose Named Views. Select V4, then select Restore. Then
    Choose OK. Your drawing should look like the following figure:

Figure 18-32: View V4

2. From the Draw menu, choose Text. Then choose Multiline Text. Enter the following
    string of options at the Command prompt:

    Command: **_mtext**

    Current text style: TEXT3. Text height: 3

    *Specify first corner: select a point to the left and above the
    view*

    *Specify opposite corner or [Height/Justify/Rotation/Style/Width]:
    select a point to the right and above the centerline*

3. Choose the Properties tab, then select TEXT5 from the Style drop-down list.

4. Choose TC from the Justification drop-down list. Then choose the Character tab.

5. Choose the Underline button. Then enter the following text:

    STANDARD CAST IRON
    MANHOLE FRAME AND COVER

6. Then choose OK. Your drawing should look like the following figure:

STANDARD CAST IRON
MANHOLE FRAME AND COVER

Figure 18-33: The completed text

Importing Text

1. From the View menu, choose Named Views. Select V5, then select Restore. Then Choose OK.

2. From the Draw menu, choose Text. Then choose Multiline Text.

3. Select a point at **380,300** for the first corner, and **500,380** for the second corner.

4. Choose Import Text. Then choose gen_note from the Open dialog box. Then choose Open. The text file is displayed in the Multiline Text Editor.

5. Then choose OK. The text is placed in the drawing. The text is an RTF file which retains the bold and underline formatting when inserted into AutoCAD.

6. From the View menu, choose Zoom. Then choose Extents.

    The drawing is now complete. Your drawing should look the following figure:

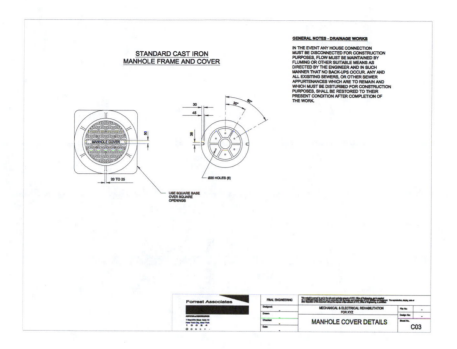

Figure 18-34: The completed drawing

7.   Use SAVEAS to save your drawing as *cover01.dwg*.

Conclusion

After completing this chapter, you have learned the following:

▶   Annotation is a an important phase in the creation of a drawing. To ensure that this phase is successful you have learned how to create new text styles using the style command. Having created new styles, you placed text in a drawing using the DTEXT command for single line text, and MTEXT for paragraph text.

▶   To edit existing text you reviewed the capabilities of the DDEDIT, and DDMODIFY commands. Using these commands you changed an existing value, created a text object with different text heights within the same word. As a final check, you performed a spell check using the spell command.

▶   Using an existing text object, you copied its properties to other text objects.

Chapter 19

Editing Text

Text can be edited in an AutoCAD drawing using either the DDEDIT or DDMODIFY commands. DDEDIT is best suited for editing the content of single line text, whereas DDMODIFY changes the content along with the text insertion point, style, justification, size, and orientation properties.

About This Chapter

In this chapter, you will do the following:

- Use control codes to place special characters.

- Place paragraph-based text using MTEXT.

- Edit existing text using DDEDIT, DDMODIFY, and MATCHPROP.

- Check text for correct spelling using SPELL.

Editing the Contents of a Text Object - Using DDEDIT

DDEDIT can be used to edit the contents of an existing text object. Methods for invoking the DDEDIT command include:

- ◗ **Toolbar:** Modify II
- ◗ **Menu:** Modify > Object > Text
- ◗ **Commands:** DDEDIT

DDEDIT lets you select text that was placed using DTEXT or MTEXT. If you do not want to retain the edits, you can choose the Undo option. DDEDIT will continue to prompt for an annotation object until you press ENTER.

Command: **_ddedit**
<Select an annotation object>/Undo:

If you select a text object that was placed using DTEXT, then the Edit Text dialog box is displayed, as shown in the following figure:

Figure 19-1: The Edit Text dialog box

If the text object was placed using MTEXT, then the Multiline Text Editor is displayed.

Editing Object Properties - Using DDMODIFY

DDMODIFY lets you edit the properties of an AutoCAD object. Methods for invoking the DDMODIFY command include:

- ◗ **Toolbar:** Standard
- ◗ **Menu:** Modify > Properties
- ◗ **Command:** DDMODIFY

Selecting an object that was placed using DTEXT will display the Modify Text dialog box, as shown in the following figure:

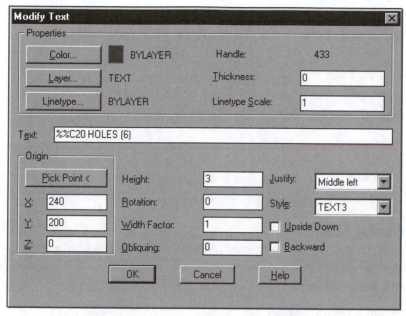

Figure 19-2: The Modify Text dialog box

Selecting an object placed using MTEXT will display the Modify MText dialog box, as shown in the following figure:

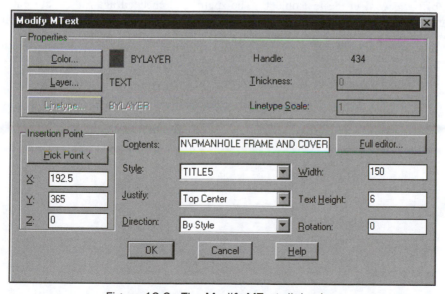

Figure 19-3: The Modify MText dialog box

The Modify MText dialog box lists the contents of the text object and a Full editor button. Selecting this button will display the Multiline Text Editor.

Using Match Properties to Assign Text Properties

The MATCHPROP command can be used to copy the properties of one object to one or more objects.

> ▶ **Toolbar:** Standard

> ▶ **Menu:** Modify > Match Properties

> ▶ **Commands:** MATCHPROP

Using Match Properties you can change the color, layer, and text properties of an Mtext object. You can also change the color, layer, linetype, linetype scale, thickness, and text properties of a text object.

Spell Checking Text

Text objects in an AutoCAD drawing can be spell checked using the SPELL command, accessed in the following ways:

> ▶ **Toolbar:** Standard

> ▶ **Menu:** Tools > Spelling

> ▶ **Commands:** SPELL

Using the spell check feature is similar to using most AutoCAD commands. First, you select the object that you want to edit using any selection set method, including all. AutoCAD will check the spelling of each word in the text object. The spelling is checked against the current main dictionary and the custom dictionary. If there are any errors, the Check Spelling dialog box is displayed, as shown in the following figure:

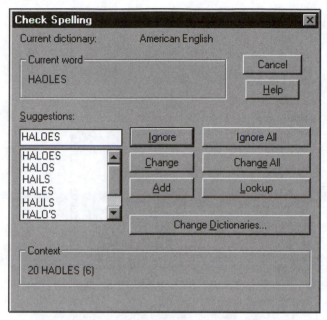

Figure 19-4: The Check Spelling dialog box

The dialog box displays the current dictionary, the current word that is misspelled, suggestions for replacement, and the context in which the misspelled word is displayed. Using this information, you have a number of choices from which to select.

⯈   You can Ignore the current word, or Ignore All remaining words which match the current word.

⯈   You can change the current word to match the word in the Suggestions box, or Change All matches in the selected text object.

⯈   If the current word is correctly spelled, you can Add it to the current custom dictionary.

⯈   You can check the spelling of the current word and display a more comprehensive list of suggestions using Lookup.

⯈   Change Dictionaries will display the Change Dictionaries dialog box, as shown in the following figure:

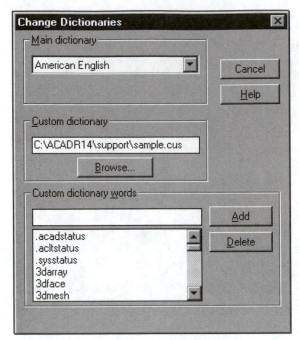

Figure 19-5: The Change Dictionaries dialog box

The Main Dictionary can be changed by selecting from the drop-down list. There is one custom dictionary supplied with AutoCAD. It is named *sample.cus* and is located in the Support directory. If your engineering discipline uses terminology not listed in the main dictionary, you can Add to *sample.cus* or create a new *.cus* file. When you have completed the spell checking, a dialog box is displayed, as shown in the following figure:

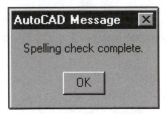

Figure 19-6: The AutoCAD Message

Exercise 19-1: Editing Text

Text objects placed in an AutoCAD drawing can be edited using either the DDEDIT, DDMODIFY, or MATCHPROP commands. In this exercise you will use these commands to make changes to existing text objects. You will also use the SPELL command to ensure that all text objects are correctly spelled.

Using DDEDIT to Change a Text Object

1.   Open the file *cov_edit.dwg*. Your drawing looks the following figure:

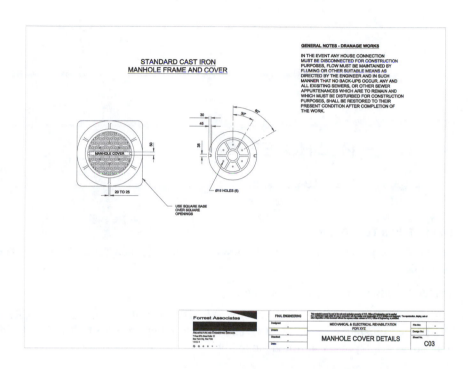

Figure 19-7: The cov_edit drawing

2.   From the View menu, choose Named Views. Select V1, then select Restore. Then choose OK.

3.   From the Modify menu, choose Object. Then choose Text.

4.   Choose the text object, Ø15 HOLES (6). The Edit Text dialog box is displayed, as shown in the following figure:

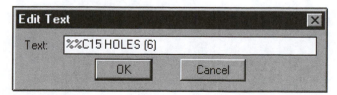

Figure 19-8: The Edit Text dialog box

5.   Change the 15 to 20, then choose OK. Your drawing should look like the following

figure:

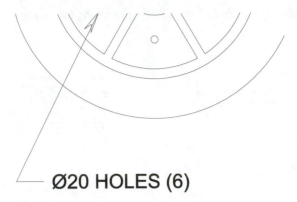

Ø20 HOLES (6)

Figure 19-9: The edited text

Using DDMODIFY to Edit a Text Object

1. From the View menu, choose Named Views. Select V2, then select Restore. Then choose OK.

2. From the Modify menu, choose Properties.

3. Choose the text object that starts with ARCHITECTURE and ENGINEERING SERVICES. Accept the selection. This text was placed using MTEXT. The Edit MText dialog box is displayed, as shown in the following figure:

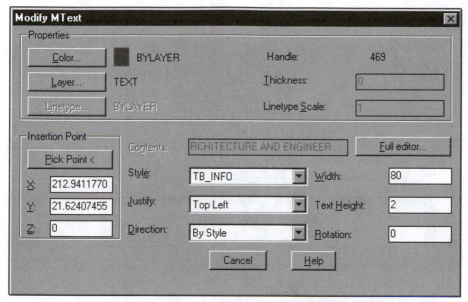

Figure 19-10: The Edit MText dialog box

9.   Choose the Full editor button. The Multiline Text Editor dialog box is displayed.

10.   Select the letter A in ARCHITECTURAL and then enter 3 in the Font Height field. Then press ENTER. Repeat this process for the first letter in Engineering and also Services. Your text is now displayed as shown in the following figure:

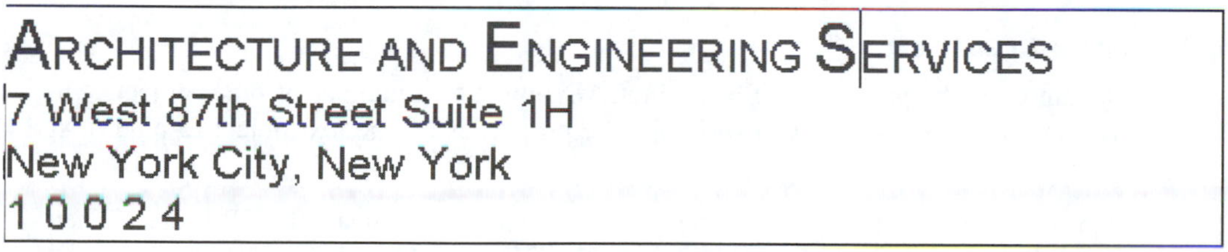

ARCHITECTURE AND ENGINEERING SERVICES
7 West 87th Street Suite 1H
New York City, New York
1 0 0 2 4

Figure 19-11: The edited text

11.   From the right-click menu, choose Select All. Then choose Red from the Text color drop-down list. The choose OK twice. Your drawing should be displayed with the change in text height, and with the color changed to yellow.

Using SPELL to Check for Spelling Errors

1.   From the cursor menu, choose Named Views. Select V3, then select Restore. Then Choose OK.

2.   From the Tools menu, choose Spelling.

3.   Choose the multiline text object. The Check Spelling dialog box is displayed, as shown

in the following figure:

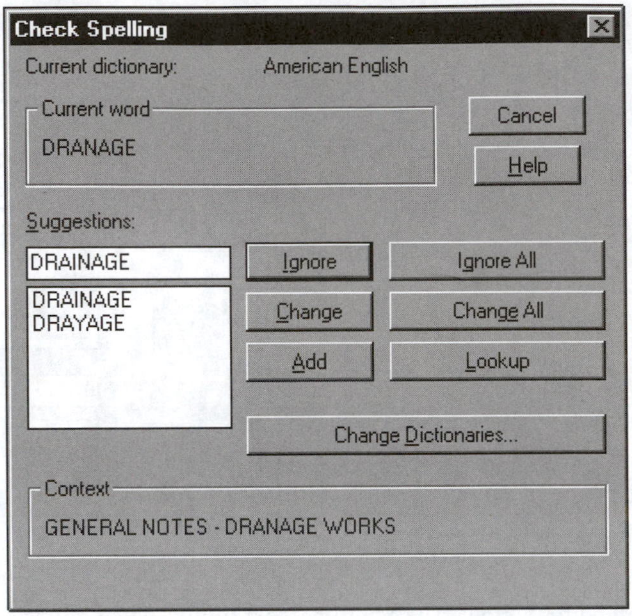

Figure 19-12: The Check Spelling dialog box

4.   Choose Change All. For all other spelling errors, choose Add. An AutoCAD Message is displayed. Then choose OK.

Using MATCHPROP to Edit Text

You can copy the properties of one AutoCAD object to a number of other objects. In this section of the exercise you will copy the properties of one text object to other text objects.

1.   From the View menu, choose Named Views. Select V4, then select Restore. Then Choose OK. Your drawing should look like the following figure:

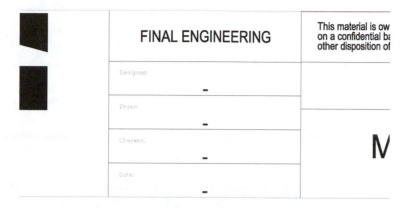

Figure 19-13: View 4

The text objects, Designed, Drawn, Checked, and Date were placed on the Hidden layer with the TEXT1 style. They should be on the Text layer, and they should be using the TB_INFO style.

2.   From the Modify menu, choose Match Properties.

Command: _matchprop

Select Source Object: *select the text object in the top right corner that starts with This material is*

Current active settings = color layer ltype ltscale thickness text dim hatch

Settings/<Select Destination Object(s)>: *using an implied window, choose the 4 text objects* Other corner: 4 found

Settings/<Select Destination Object(s)>:

The properties are copied and the 4 text objects are displayed on the correct layer, and with the correct text style.

3.   Use Saveas to save the drawing as *cover03.dwg.*

Conclusion

After completing this chapter, you have learned the following:

▶   To edit existing text you reviewed the capabilities of the ddedit, and ddmodify commands. Using these commands you changed an existing value, created a text object with different text heights within the same word. As a final check, you performed a spell check using the spell command.

▶   Using an existing text object, you copied its properties to other text objects.

Chapter 20

Placing Dimensions

In this chapter, you learn how to place dimensions that conform to local and national standards. Dimensioning is a critical component in the creation of a working drawing. When combined with the necessary views and annotations, dimensions provide a complete set of instructions for the construction of a part, assembly, or building.

About This Chapter

In this chapter, you will do the following:

- ▶ Review dimensioning terminology.

- ▶ Place dimensions using various DIM commands.

- ▶ Annotate a component using the LEADER command.

- ▶ Add dimensions which include tolerances.

- ▶ Edit existing dimensions.

- ▶ Create new dimension styles using the DDIM command.

Dimension Terminology

When you dimension an object in AutoCAD you use extension lines, dimension lines, leaders, arrowheads, annotation, and symbols. AutoCAD can automatically generate these based on the settings you select from Geometry, Format, and Annotation dialog boxes accessed from the Dimension Style dialog box, as shown in the following figure:

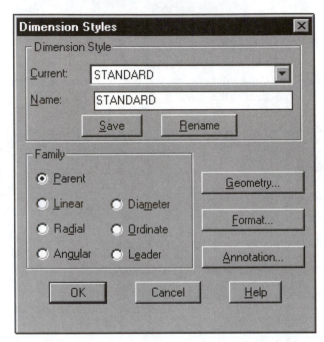

Figure 20-1: The Dimension Styles dialog box

AutoCAD supplies one style, STANDARD, with the *acad.dwt* template file. This style contains settings that are considered typical for most applications using imperial units. If you use the *acadiso.dwt* template file, the settings in STANDARD are modified to suit metric applications. There are also dimension styles provided for DIN, ISO, and JIS applications. You can access these styles by using the appropriate template file from the Use a Template, or when you select a title block in the Advanced Setup Wizard. Using the STANDARD style, a small mechanical object would be dimensioned, as shown in the following figure:

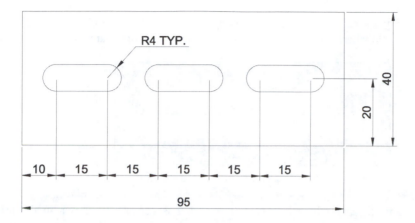

Figure 20-2: Using the Standard dimension style

A typical linear dimension, as shown in the previous figure, consists of settings that define the dimension line, extension lines, arrowheads, and text. You must also set the dimension line offset, and the extension line offset, as shown in the following figure:

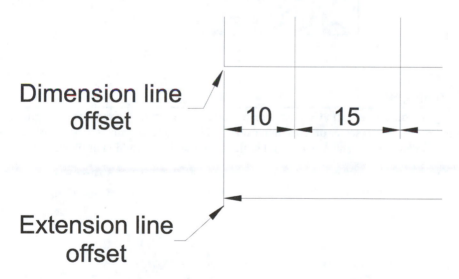

Figure 20-3: Typical dimension line terminology

Dimension Styles Dialog Box

To check on the current values of these settings or to change the settings, you must become familiar with the contents of the Dimension Styles dialog box.

The Dimension Style area, shown in the following figure, displays the Current style name, and provides a Name field where you can create and save a new style, or rename an existing style.

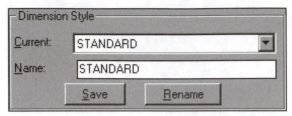

Figure 20-4: The Dimension Style section

The Family section displays which Family style member is current, as shown in the following figure.

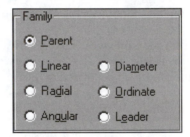

Figure 20-5: The Family style section

Geometry Dialog Box

The Geometry button displays the Geometry dialog box. The settings control the appearance of the geometry and overall scale of the dimension. These settings include dimension lines, extension lines, arrowheads, center marks, and scale, as shown in the following figure:

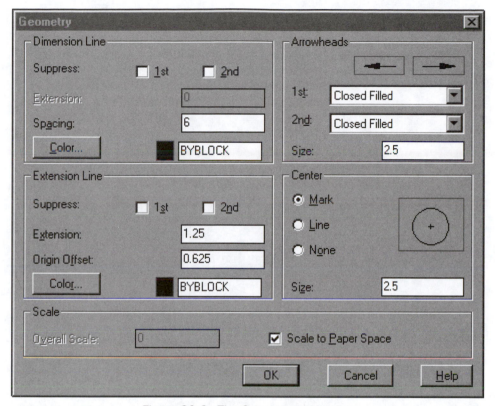

Figure 20-6: The Geometry dialog box

The Dimension Line section lets you suppress one or both dimension lines. The selection of which line to suppress is based upon the order in which you dimension the object, as shown in the following figure:

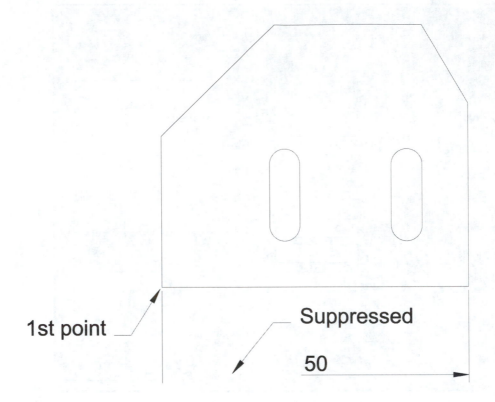

Figure 20-7: Suppressing the 1st dimension line

The Extension setting is displayed when you select oblique, architectural tick, or integral for the arrowhead style. The value you enter specifies the distance to extend the dimension line beyond the extension line, as shown in the following figure:

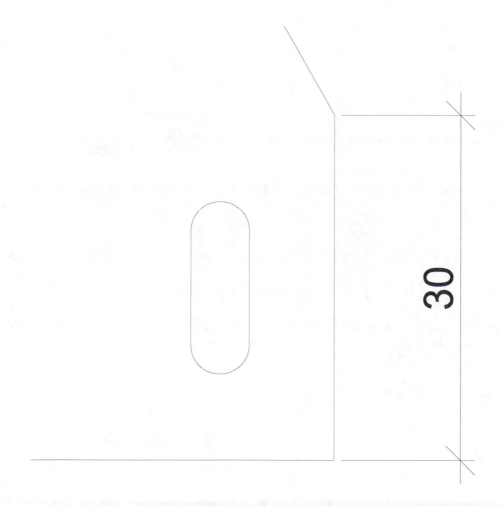

Figure 20-8: The extended dimension line

Spacing sets the distance between the dimension lines of a baseline dimension, as shown in the following figure:

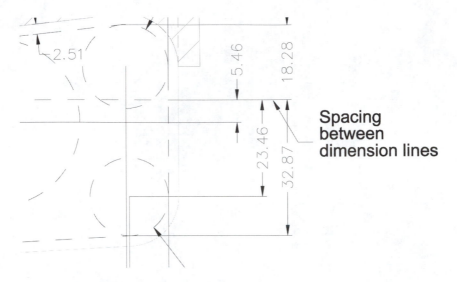

Figure 20-9: Spacing between dimension lines

Using the Color option you can display the dimension line in a different color than the extension lines and text. Choosing either the Color button or the color swatch will display the Select Color dialog box, as shown in the following figure:

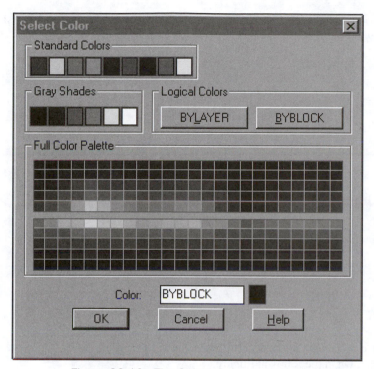

Figure 20-10: The Select Color dialog box

The Extension Line area of the dialog box lets you suppress one or both extension lines. The selection of which line to suppress is based upon the order you dimension the object, as shown in the following figure:

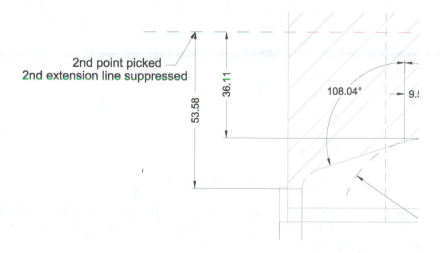

Figure 20-11: Suppressing the 2nd extension line

The Extension option sets the distance the extension line extends above the dimension line, as shown in the following figure:

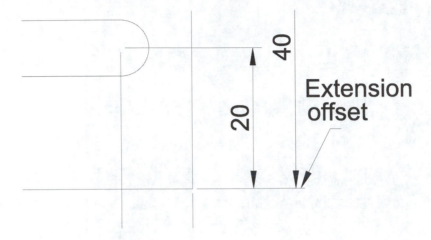

Figure 20-12: The extended extension line

The Origin Offset option sets the distance to offset the extension line from the points that define the dimension, as shown in the following figure:

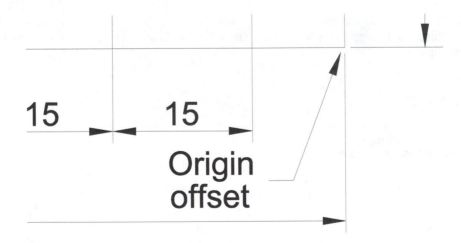

Figure 20-13: The offset distance

Using the Color option you can display the extension line in a different color from the dimension lines, and text. Choosing either the Color button or the color swatch will display the Select Color dialog box, as shown in Figure 20-10.

The Arrowheads area controls the appearance of the arrowheads, as shown in the following figure:

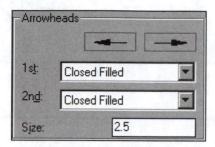

Figure 20-14: The Arrowhead area

When you select an arrowhead, the style is displayed in the image tile. The 1st and 2nd arrowhead can be set to different values by selecting the appropriate style from the 1st and 2nd drop-down lists. You can set the size of the arrowheads in the Size field.

The Center area of the dialog box sets the appearance of center marks and centerlines for diametric and radial dimensions. The Mark option creates a center mark based upon the value in the Size field, as shown in the following figure:

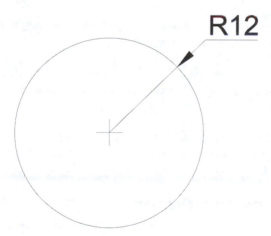

Figure 20-15: A center mark

The Line option creates a center line if the value set in the Size field is set to a negative value, as shown in the following figure:

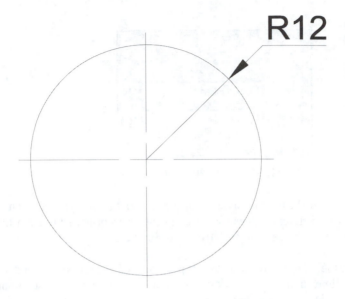

Figure 20-16: A center line

The None option creates no center line and sets the value in the Size field to 0. The Size field sets the size of the center mark and center line as noted above.

In the Scale area, you can set a scale factor for all dimension settings that set size distance or spacing. Text and arrowhead sizes are affected, but tolerances, measured lengths, coordinates, and angles are not affected.

If you set Scale to Paper Space on, AutoCAD determines a scale factor based on the scaling between the current model space viewport and paper space. This setting maintains the size of your dimensions relative to the object even when the drawing is scaled in paper space.

Placing Dimensions on a Drawing

Dimensioning an object correctly requires a number of different dimensioning methods based on the object you are dimensioning. This includes the linear method for horizontal and vertical dimensions, aligned dimensions, the angular method for angles, the diameter and radius methods for circles and arcs, and the ordinate method for datum-based dimensions. You can also place multiple dimensions using the baseline or continuous options.

Linear

The DIMLINEAR command places a dimension in a horizontal, vertical, or rotated position. To place a dimension, you can select 2 points which define the extension line origin, or you can press ENTER to select an object.

Methods for invoking the DIMLINEAR command include:

▶ **Toolbar:** Dimension

▶ **Menu:** Dimension > Linear

▶ **Command:** DIMLINEAR

When you place a dimension by selecting 2 points, the direction in which you move the cursor determines whether the dimension is horizontal or vertical, as shown in the following figure:

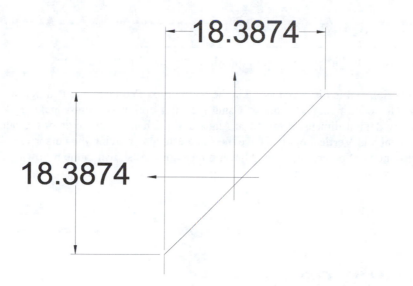

Figure 20-17: Placing a horizontal or vertical dimension

As you create the dimension you can modify the text and the angle of the text, or rotate the angle of the dimension line. The Command prompt text looks similar to the following:

```
Command: _dimlinear
First extension line origin or press ENTER to select:
Second extension line origin:
Dimension line location (Mtext/Text/Angle/Horizontal/Vertical/Rotated):
Dimension text = 125.000
```

To modify the text you can enter the **Mtext** option. This will display the Multiline Text Editor dialog box, as shown in the following figure:

Figure 20-18: The Dimension text in the Multiline Text Editor dialog box

The default dimension text is displayed as <> in the dialog box. You can add text before or after the <>, and use Unicode or the %% codes. Using the Text option you must include <> if you want to include the default dimension text. The Angle option changes the angle of the dimension text. The Horizontal and Vertical options have the same effect as moving the cursor in a specific direction. The Rotated option creates rotated linear dimensions, as shown in the following figure:

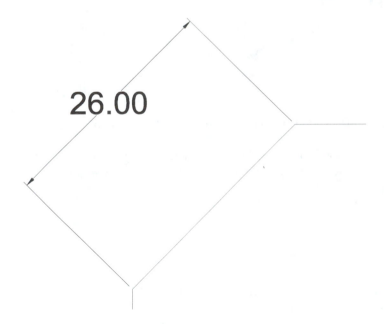

Figure 20-19: A rotated linear dimension

Aligned

Aligned dimensions are parallel to the extension line origins, and they can be created using the DIMALIGNED command. Methods for invoking the DIMALIGNED command include:

> ▶ **Toolbar:** Dimension

> ▶ **Menu:** Dimension > Aligned

> ▶ **Command:** DIMALIGNED

To place an aligned dimension you can select 2 points which define the extension line origin, or you can press ENTER to select an object, as shown in the following figure:

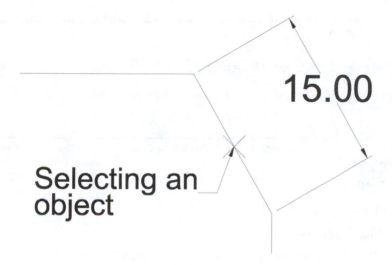

Figure 20-20: An aligned dimension

You create an aligned dimension and modify the default dimension text using the same techniques as the DIMLINEAR command. You can modify the text using the Mtext, Text, or Angle Command prompt options.

Ordinate

Ordinate or datum dimensions display the X or Y ordinate of a feature. The dimension is placed along with a simple leader line. Ordinate dimensions can be created using the DIMORDINATE command. Methods for invoking the DIMORDINATE command include:

> ▶ **Toolbar:** Dimension

> ▶ **Menu:** Dimension > Ordinate

> ▶ **Command:** DIMORDINATE

To place an ordinate dimension you select the feature you want to dimension. The current UCS is used to measure the X or Y ordinate. The leader is placed in an orthogonal direction to the UCS, as shown in the following figure:

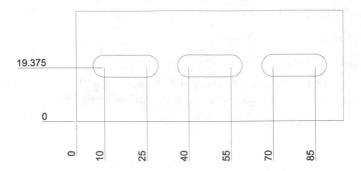

Figure 20-21: Creating ordinate dimensions

You can edit the text using the MText or Text Command prompt options.

Radius

The DIMRADIUS creates radial dimensions for circles and arcs. Methods for invoking the DIMRADIUS command include:

▶ **Toolbar:** Dimension

▶ **Menu:** Dimension > Radius

▶ **Command:** DIMRADIUS

To place a dimension on an arc or circle you select the object. The position of the cursor determines the location of the dimension text, as shown in the following figure:

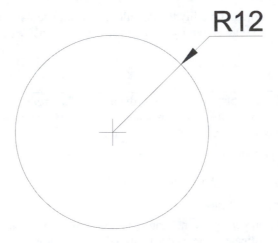

Figure 20-22: Dimensioning an arc or circle

You can modify the text using the Mtext, Text, or Angle Command prompt options.

Centermark

When you place a dimension using either the DIMRADIUS or DIMDIAMETER commands, you can control whether or not a center mark or center lines are included. This can be done in the Geometry dialog box accessed from the Dimension Styles dialog box, or with the DIMCEN system variable. The values for DIMCEN are as shown in the following:

> Note: This can also be done using the DIMCEN system variable. The values for DIMCEN are as follows:

- 0 - No center marks or lines are drawn

- <0 - Centerlines are drawn

- >0 - Center marks are drawn

The size of the center marks and center lines must be set in the Geometry dialog box. The center mark will only be placed if your dimension is outside of the arc or circle.

Diameter

The DIMDIAMETER command creates diameter dimensions for circles and arcs. Methods for invoking the DIMDIAMETER command include:

- **Toolbar:** Dimension

- **Menu:** Dimension > Diameter

- **Command:** DIMDIAMETER

To place a dimension on an arc or circle, select the object. The position of the cursor determines the location of the dimension text, as shown in the following figure:

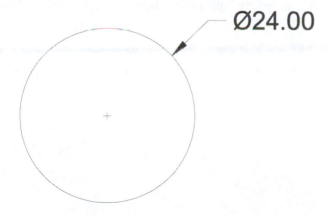

Figure 20-23: Placing a diameter dimension

Angular

The DIMANGULAR command creates an angular dimension on an arc, circle or line.

Methods for invoking the DIMANGULAR command include:

- ▶ **Toolbar:** Dimension
- ▶ **Menu:** Dimension > Angular
- ▶ **Command:** DIMANGULAR

To place an angular dimension, you can select either an arc, circle, or line as shown in the following figure:

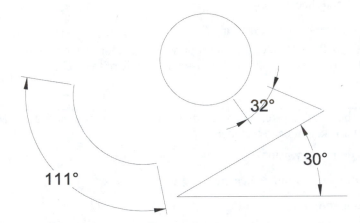

Figure 20-24: Placing angular dimensions

You can also press ENTER to specify 3 points that define the angle, as shown in the following figure:

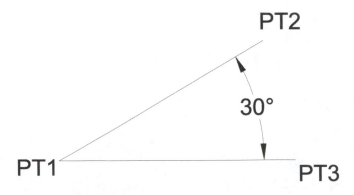

Figure 20-25: Using 3 points

Baseline

The DIMBASELINE command lets you continue a linear, angular, or ordinate dimension from the baseline of the previous or selected dimension. Methods for invoking the DIMBASELINE command include:

▶ **Toolbar:** Dimension

▶ **Menu:** Dimension > Baseline

▶ **Commands:** DIMBASELINE

When you place baseline dimensions you can use the most recently-placed linear, angular, or ordinate dimension as the starting point, as shown in the following figure:

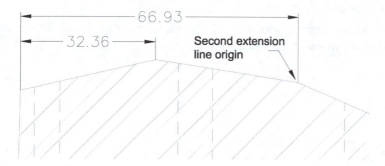

Figure 20-26: Specifying a second extension line origin

You can also select a previously-placed dimension, as shown in the following figure:

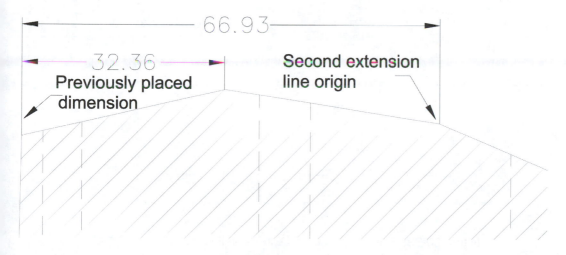

Figure 20-27: Selecting a base dimension

The spacing between each dimension can be set in the Geometry dialog box in the Spacing field.

> Note: You can also set dimension spacing with the DIMDLI system variable.

Continue

The DIMCONTINUE command lets you continue a linear, angular, or ordinate dimension from the second extension line of the previous or selected dimension.

Methods for invoking the DIMCONTINUE command include:

▶ **Toolbar:** Dimension

▶ **Menu:** Dimension > Continue

▶ **Command:** DIMCONTINUE

When you continue dimensions you can use the most recently-placed linear, angular, or ordinate dimension as the starting point.

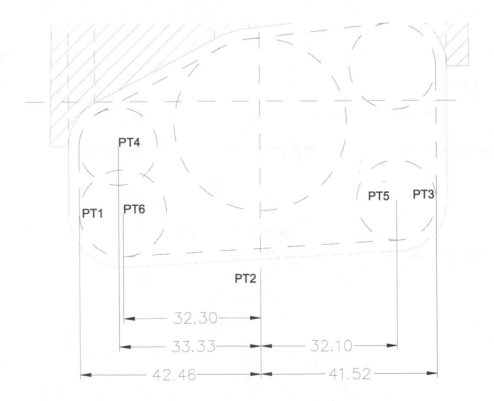

Figure 20-28: Specifying a second extension line origin

You can also select a previously-placed dimension. This is similar to the process for baseline dimensioning.

Leader

A leader connects an annotation to a feature in your drawing. You can place a leader by using the LEADER command. Methods for invoking the DIMLEADER command include:

> ▶ **Toolbar:** Dimension

> ▶ **Menu:** Dimension > Leader

> ▶ **Commands:** LEADER

When you place a leader you can attach it to any object. The leader consists of an optional arrowhead which you set using the Format option, then the Arrow or the None option. The arrowhead selection is connected to a straight line segment or a smooth spline which is set in the Format option, by entering **Spline** or **Straight**. The annotation is an object which can be single-line or paragraph text, as shown in the following figure:

Figure 20-29: Leader lines

Tolerance

The TOLERANCE command lets you place geometric tolerances in your drawing. Methods for invoking the DIMLINEAR command include:

> ▶ **Toolbar:** Dimension

> ▶ **Menu:** Dimension > Tolerance

> ▶ **Commands:** TOLERANCE

When you place geometric tolerances in your drawing, a feature control frame is created. This frame consists of a number of compartments which display geometric characteristic symbols followed by one or more tolerance values. Where appropriate, the compartments also display a diameter symbol with datums and their material condition, as shown in the following figure:

Figure 20-30: Feature control frame

Associative Dimensions

When you place a dimension in AutoCAD it can be drawn with all the lines, arrowheads, arcs, and text as a single dimension object. This is controlled by the DIMASO system variable. If DIMASO is on, the dimension is a single object. When you edit an edit an object with associative dimensions and DIMASO on, the dimensions will be updated. If it is off, each part of the dimension is drawn as a single object and updating of the dimensions does not occur.

Exercise 20-1: Creating Dimensions

 When you dimension a drawing you should plan what will be the most effective method of placing the dimensions. Changing the settings in the DDIM dialog box, and using features such as baseline and continuous dimensions will improve your productivity.

In this exercise you will use the Standard dimension style to place linear, radial, leader, baseline, and continuous dimensions.

Using dimlinear to Place Dimensions

1. From the File menu, choose Open. Select *dimens1.dwg* then choose Open. Your drawing should look like the following figure:

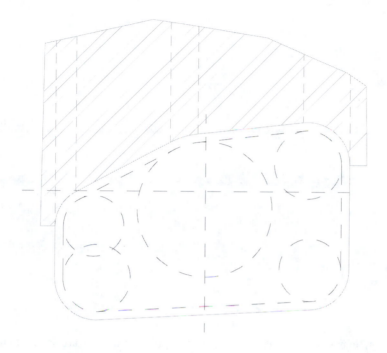

Figure 20-31: Dimens1

2.   From the View menu, choose Named Views. The View Control dialog box is displayed. Then choose V1. Choose the Restore button, then choose OK. Your drawing should look like the following figure:

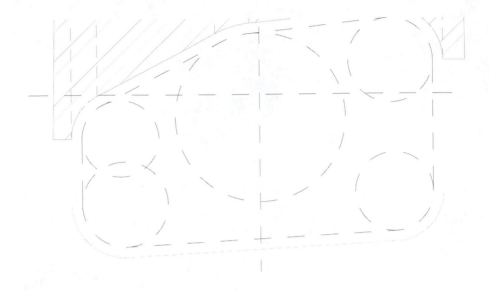

Figure 20-32: View V1

3. From the Dimension menu, choose Linear. Endpoint and center are set as running object snaps. Place the dimension.

```
Command: _dimlinear
```

First extension line origin or press ENTER to select: *select the endpoint at PT1 as shown in the following figure*

Second extension line origin: *select the endpoint at PT2 as shown in the following figure*

Dimension line location (Mtext/Text/Angle/Horizontal/Vertical/Rotated): *select a point approximately 25 units from the drawing*

```
Dimension text = 42.46
```

Your drawing should look like the following figure:

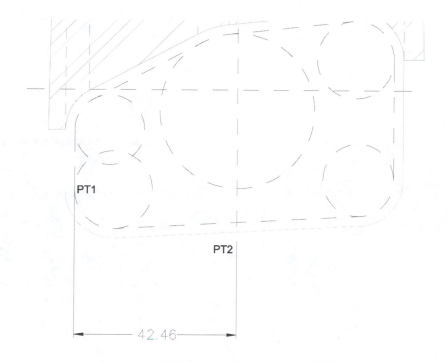

Figure 20-33: The linear dimension

4.   From the Dimension menu, choose Continue. Place the dimension.

Command: _dimcontinue

Specify a second extension line origin or (Undo/<Select>): *select the endpoint of the line at PT3 as shown in the following figure*

Dimension text = 41.52

Specify a second extension line origin or (Undo/<Select>):

Select continued dimension: *press ENTER to complete the command*

Your drawing should look like the following figure:

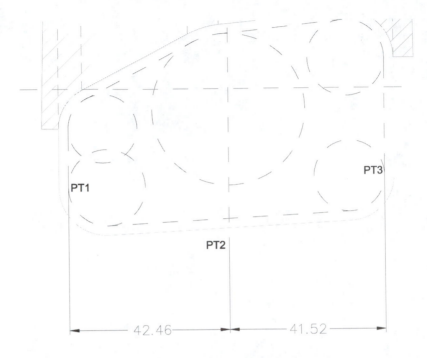

Figure 20-34: The continue dimension

5. From the Dimension menu, choose Style. Then choose Geometry. In the Extension Line section of the dialog box select Suppress 2nd, as shown in the following figure:

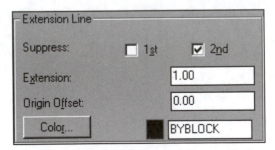

Figure 20-35: The Extension Line settings

6. Then choose OK twice to exit.

7. From the Dimension menu, choose Linear. Place the dimension by selecting the center of the circle at PT4 and the endpoint of the line at PT2 as shown in the following figure:

Place the dimension approximately 6 units above the 42.46 dimension.

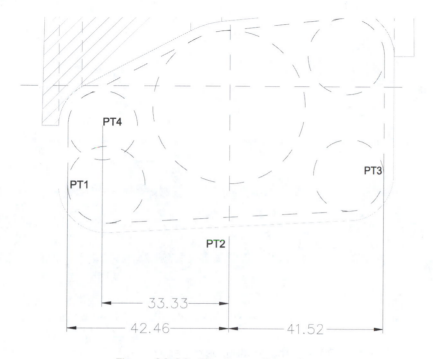

Figure 20-36: The linear dimension

8. From the Dimension menu, choose Style. Then choose Geometry. In Extension Line ensure that neither line is suppressed. Then choose OK twice.

9. From the Dimension menu, choose Continue. Place the dimension by selecting the center of the circle at PT5 as shown in the following figure:

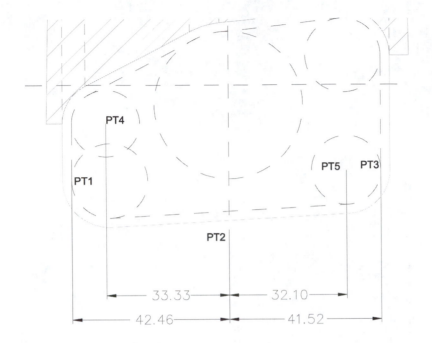

Figure 20-37: The continuous dimension

10. Repeat Step 5.

11. From the Dimension menu, choose Linear. Place the dimension by selecting the center of the circle at PT6 and the endpoint of the line at PT2 as shown in the following figure:

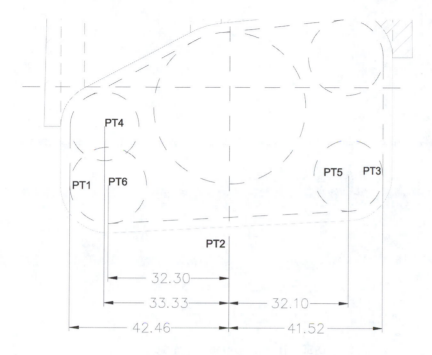

Figure 20-38: The linear dimension

12.   Repeat Step 8.

13.   From the Dimension menu, choose Continue. Place the dimension by selecting the center of the circle at PT7 as shown in the following figure:

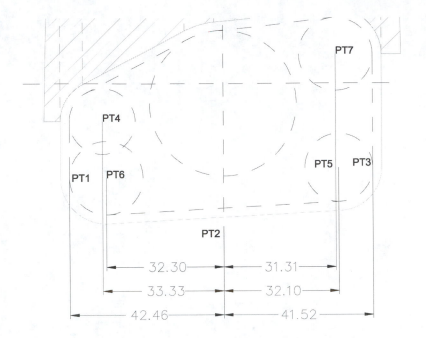

Figure 20-39: The continued dimension

Placing Radius Dimensions

1. From the Dimension menu, choose Radius. Place the dimension by choosing the circle and locating the dimension as shown in the following figure:

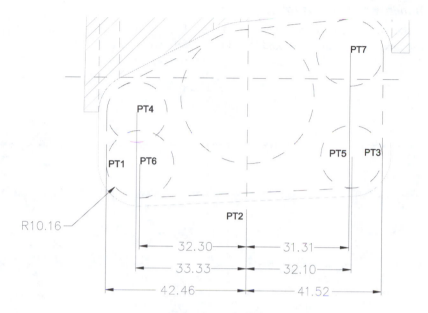

Figure 20-40: The radius dimension

2.   Repeat Step 1 and dimension the 2 circles as shown in the following figure:

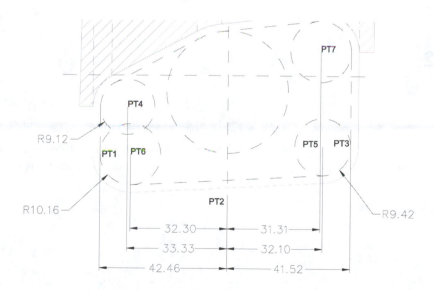

Figure 20-41: The completed radius dimensions

Adding more Dimensions to the Drawing

1.   From the View menu, choose Named Views. Then choose V2. Choose Restore, then choose OK. Your drawing should look like the following figure:

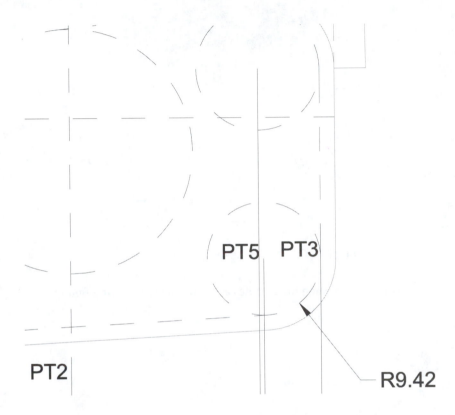

Figure 20-42: View V2

2.   From the Dimension menu, choose Style. Then choose Format. In the Text section, ensure that neither line Inside Horizontal or Outside Horizontal is selected as shown in the following figure:

Then choose OK twice.

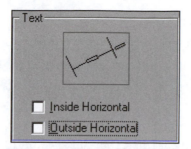

Figure 20-43: The Text settings

3.   From the Dimension menu, choose Linear. Place the dimension by selecting the center
      of the circle at PT8 and the endpoint of the line at PT9 as shown in the following figure:

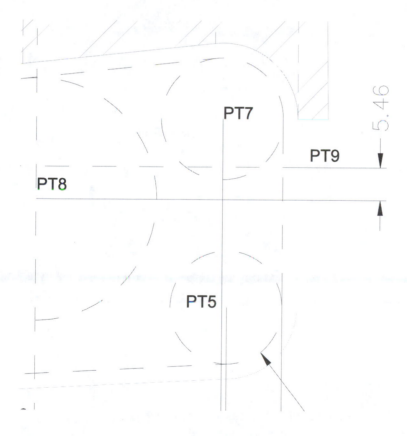

Figure 20-44: The linear dimension

4.   From the Dimension menu, choose Baseline. Place the dimension.

```
Command: _dimbaseline
```

Specify a second extension line origin or (Undo/<Select>): **s**

Select base dimension: *select the extension line closest to PT9*

Specify a second extension line origin or (Undo/<Select>): *select the center of the circle at PT5*

Dimension text = 23.46

Specify a second extension line origin or (Undo/<Select>):

Select base dimension:

Your drawing should look like the following figure:

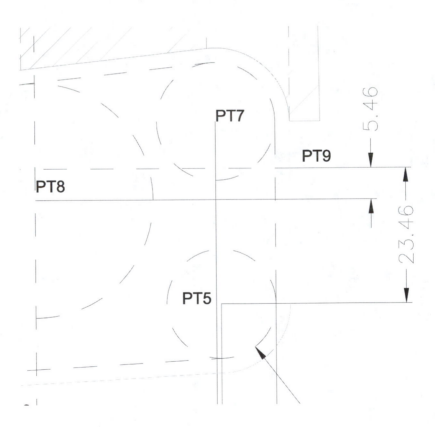

Figure 20-45: The baseline dimension

5. From the Dimension menu, choose Baseline. Place the dimension.

```
Command: _dimbaseline
```

Specify a second extension line origin or (Undo/<Select>): *select the endpoint of the line at PT10*

Dimension text = 32.88

Specify a second extension line origin or (Undo/<Select>):

Select base dimension:

Your drawing should look like the following figure:

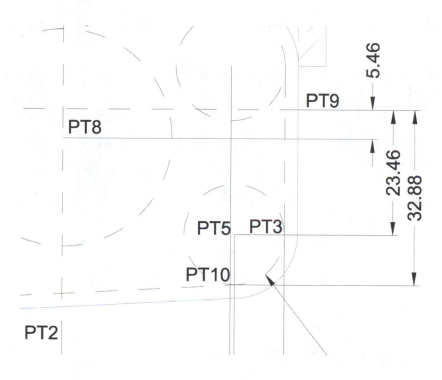

Figure 20-46: The baseline dimension

6.   From the Dimension menu, choose Continue. Place the dimension.

Command: _dimcontinue

Specify a second extension line origin or (Undo/<Select>): **s**

Select continued dimension: *select the extension line at PT9*

Specify a second extension line origin or (Undo/<Select>): *select
the endpoint of the line at PT11*

Dimension text = 18.28

Specify a second extension line origin or (Undo/<Select>):

Select continued dimension:

Your drawing should look like the following figure:

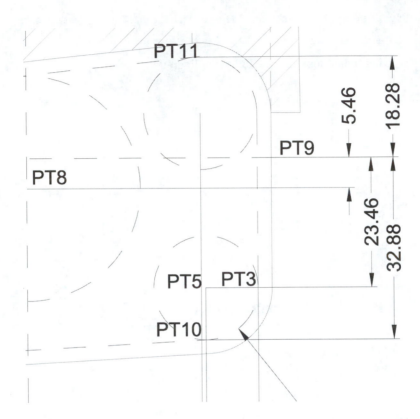

Figure 20-47: The continue dimension

Using DIMALIGNED to place dimensions

1. From the View menu, choose Named Views. Then choose V3. Choose Restore, then choose OK. Your drawing should look like the following figure:

Figure 20-48: View V3

2.   From the Dimension menu, choose Style. Then choose Format. In the Text section ensure that Outside Horizontal is on. A checkmark indicates that it is on. Then choose OK.

3.   Then choose Geometry. Suppress the 1st and 2nd extension lines. Then choose OK twice.

4.   Toggle OSNAP on the Status toolbar to off.

5.   From the Dimension menu, choose Aligned. Place the dimension.

Command: _dimaligned

First extension line origin or press ENTER to select: _nea to
select a point close to PT12

Second extension line origin: _per to select a point close to
PT13

Dimension line location (Mtext/Text/Angle):

Dimension text = 2.51

Your drawing should look like the following figure:

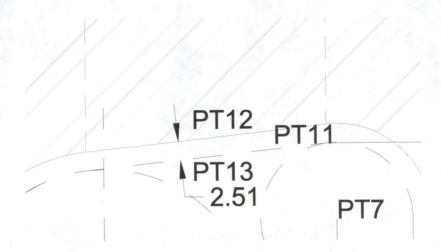

Figure 20-49: The aligned dimension

6. Using the techniques from this exercise complete the dimensioning of the part. The completed drawing should look like the following figure:

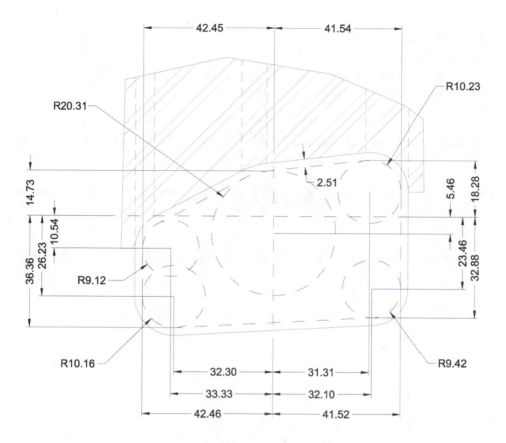

Figure 20-50: The completed drawing

7.   From the File menu, choose SaveAs. Save your drawing using a suitable filename.

Conclusion

After completing this chapter, you have learned the following:

▶ A dimension consists of extension lines, dimension lines, leaders, arrowheads, annotation, and symbols.

▶ You can create new dimension styles, and set the appropriate values in the Geometry, Format, and Annotation sections.

▶ You have learned the procedure for placing annotation on a drawing using a leader line.

▶ You can create suitable values for placing tolerance dimensions.

Chapter 21

Editing Dimensions and Creating Dimension Styles

About This Chapter

In this chapter, you will do the following:

- ◗ Edit existing dimensions.

- ◗ Create new dimension styles using the DDIM command.

Editing Dimensions

After you have placed a dimension in a drawing, you may want to modify the dimension. You can do this using the AutoCAD edit commands and grip editing modes.

When you place a dimension, AutoCAD places *definition points* to determine the location of the dimensions. These points are placed on a layer named DEFPOINTS which is created by AutoCAD. The definition points are not plotted. The importance of these points in editing is that they must be included in the selection set if the dimension update is to work correctly. Grip points are located at the definition points, making the use of the grip editing capabilities generally easier and more effective.

> Note: The definition points should not be edited or erased.

The DIMEDIT Command

To edit dimensions, use the DIMEDIT command. Methods for invoking the DIMEDIT command include:

- ▶ Toolbar: Dimension
- ▶ Menu: Dimension > Oblique
- ▶ Commands: DIMEDIT

You are given four options at the Command prompt:

- ▶ *Home* - returns the dimension text back to its default position
- ▶ *New* - lets you change the dimension text using the Multiline Text Editor dialog box
- ▶ *Rotate* - rotates the dimension text
- ▶ *Oblique* - applies to extension lines. AutoCAD creates extension lines perpendicular to the dimension line. The Oblique option lets you adjust this angle if the extension lines are interfering with other features in your drawing.

The DIMTEXT Command

Dimension text can be rotated, replaced with existing text, or moved to a new location. The original location of the text defined in the text style is named the home position, and can be used to return the text to its original location.

Methods for invoking the DIMTEDIT command include:

- ▶ **Toolbar:** Dimension
- ▶ **Menu:** Dimension > Align Text
- ▶ **Commands:** DIMTEDIT

You are given four options at the Command prompt:

- ▶ *Left* - left-justifies the text on the dimension line. This option is available for linear, radial, and diametric dimensions.

▶ *Right* - right-justifies the text on the dimension line. This options is available for linear, radial, and diametric dimensions.

▶ *Home* - returns the dimension text back to its default position

▶ *Angle* - changes the angle of the dimension text

Dimension Overrides

In most cases you will create dimension styles to place dimensions in your drawing. These styles will set the necessary variables according to local and national standards. When you want to make a minor change to an individual dimension, you can override the current style without making a permanent change to the style. This can be particularly useful in situations where you want to suppress the 1st or 2nd extension line.

To override a dimension there are a number of options you can select. The DDIM command allows you to change the options for Geometry, Format, and Annotation. A change made to the current style is indicated by the plus sign before the style name, as in +STANDARD.

The DDMODIFY command can also be used to set up overrides for an existing style. Existing dimensions can be edited using the DDMODIFY command. When you select the dimension the Modify Dimension dialog box is displayed.

The DIMOVERRIDE command prompts you for the name of the dimension variable. Options in accessing the DIMOVERRIDE command include:

▶ **Toolbar:** None

▶ **Menu:** Dimension > Override

▶ **Commands:** DIMOVERRIDE

A comprehensive listing and description of the dimension variables can be viewed in the *Release 14 Command Reference.*

Using Match Properties to Assign Dimension Properties

The MATCHPROP command lets you copy the properties from one object to one or more objects. Dimension, leader, and tolerance objects can be used with this command.

Exercise 21-1: Editing Dimensions

When you initially place dimensions on an object you may not foresee a conflict or an incorrect placement of a dimension. In this exercise you will edit existing dimensions using the DIMEDIT, DIMTEDIT, and DDMODIFY commands. You will also stretch the object and see how associative dimensions are modified automatically.

1. From the File menu, choose Open. Select *dimens2.dwg* then choose Open. Your drawing should look like the following figure:

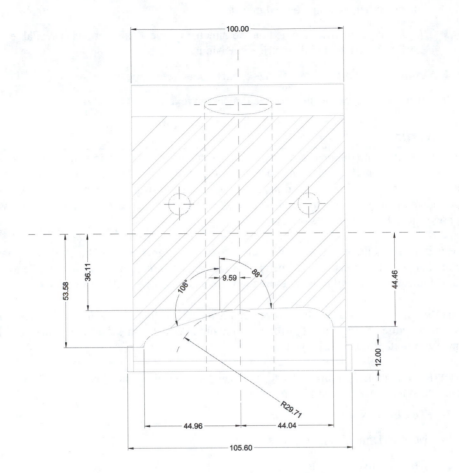

Figure 21-1: The dimens2 drawing

2. From the View menu, choose Named Views. Then choose V1. Choose Restore, then choose OK. Your drawing should look like the following figure:

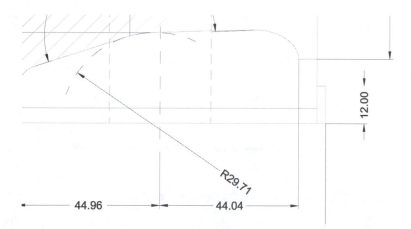

Figure 21-2: View V1

3. From the Dimension menu, choose Style. Then choose Geometry. Set the Extension Line settings as shown in the following figure:

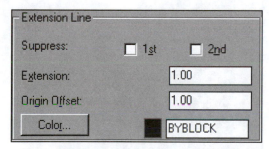

Figure 21-3: The Extension Line settings

4. Then choose OK twice.

5. From the Dimension menu, choose Update. Choose the 12.00 dimension. Your drawing should look like the following figure:

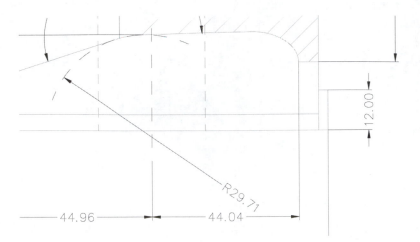

Figure 21-4: The edited dimension

6.   From the Dimension menu, choose Style. Then choose Format. Select the Text settings, as shown in the following figure:

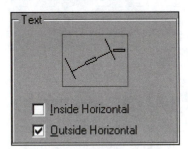

Figure 21-5: The Text settings

7.   From the Dimension menu, choose Update. Choose the R29.71 dimension. Your drawing should look like the following figure:

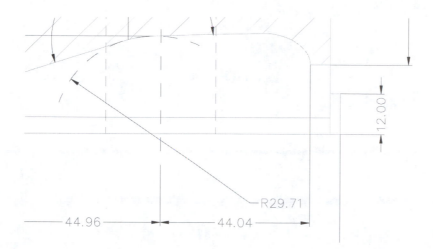

Figure 21-6: The edited dimension

8. Select the R29.71 dimension to display the grips. Make the R29.71 the hot grip and stretch the dimension to a new location, as shown in the following figure:

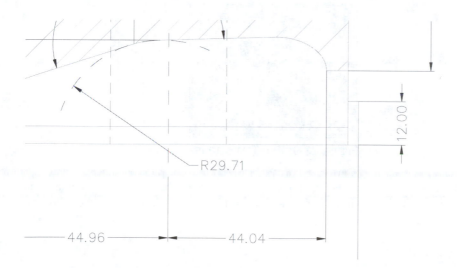

Figure 21-7: The relocated dimension

Editing Angular dimensions using grips

1. From the View menu, choose Named Views. Then choose V2. Choose Restore, then choose OK. Your drawing should look like the following figure:

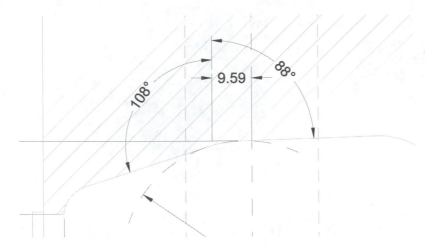

Figure 21-8: View V2

2. From the Dimension menu, choose Style. Then choose Format. Select the Text settings, as shown in the following figure:

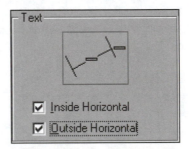

Figure 21-9: The Text settings

3. From the Dimension menu, choose Update. Choose the 108 and 87 angular dimensions. Press ENTER to exit the command. Your drawing should look like the following figure:

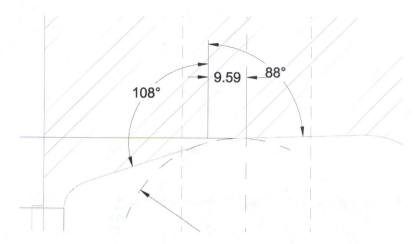

Figure 21-10: The edited angular dimensions

4.   Using grips with OSNAP toggled off, stretch the angular dimensions until they look like the following figure:

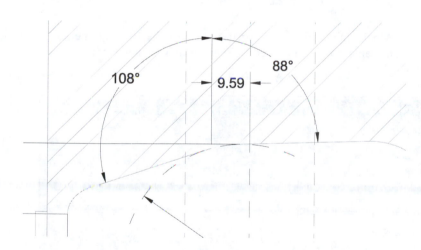

Figure 21-11: The stretched dimensions

Modifying Text Properties

1.   From the View menu, choose Named Views. Then choose V3. Choose Restore, then choose OK. Your drawing should look like the following figure:

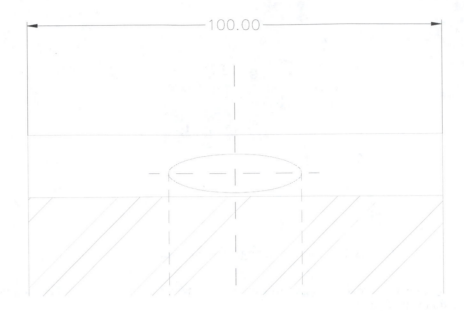

Figure 21-12: View V3

2.   From the Modify menu, choose Properties. Choose the 100.00 dimension. The Modify Dimension dialog box is displayed. Complete the Contents sections as shown in the following figure:

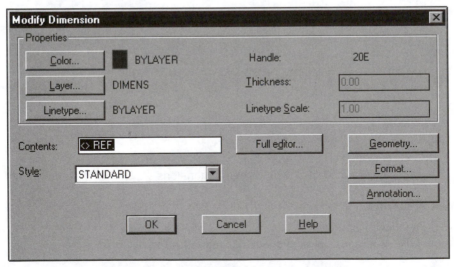

Figure 21-13: The Modify Dimension dialog box

3.   Then choose OK. Your drawing should look like the following figure:

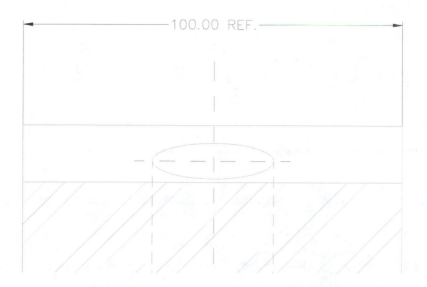

Figure 21-14: The modified dimension

Using the STRETCH Command

1. From the View menu, choose Named Views. Then choose V4. Choose Restore, then choose OK. This will display the complete drawing.

2. From the Modify menu, choose Stretch. Place a crossing window as shown in the following figure:

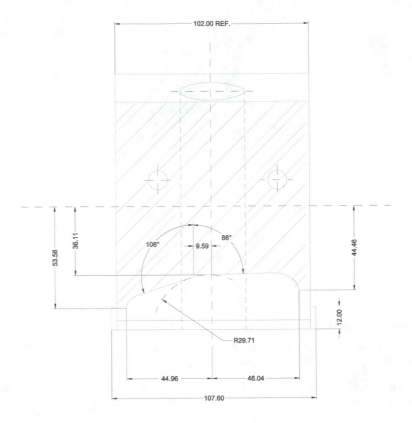

Figure 21-15: The crossing window

3.   Complete the command as shown in the following:

Command: _stretch

Select objects to stretch by crossing-window or crossing-polygon...

Select objects: Other corner: 44 found

Select objects:

Base point or displacement: *with OSNAP toggled on, choose the endpoint of the centerline near the 100.00 REF dimension*

Second point of displacement: *with ORTHO toggled on, move the cursor to the right, then enter* **2**

Your drawing will be stretched 2 units to the right. All dimensions will be updated accordingly since they are associative dimensions. Zoom and pan around the drawing to review the stretched dimensions.

The completed drawing should look like the following figure:

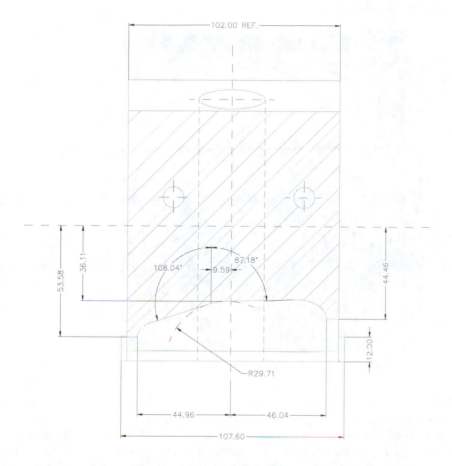

Figure 21-16: The completed drawing

4.   From the File menu, choose SaveAs. Save your drawing using a suitable filename.

Using Dimension Styles

When you dimension a drawing you can create overrides to accommodate minor differences in settings. You can also create dimension styles which match the requirements of your current project. This is another technique for maintaining standards across many drawings in a project. The dimension styles can be saved in template files and when you start a new drawing using a template file, the appropriate dimension styles will be available. Dimension styles are created using the DDIM command. The settings for Geometry, Format, and Annotation can be changed to suit the specific requirements of your dimension.

Creating a Dimension Style

To create a dimension style, you name and save a parent dimension style based on an existing style as shown in the following figure:

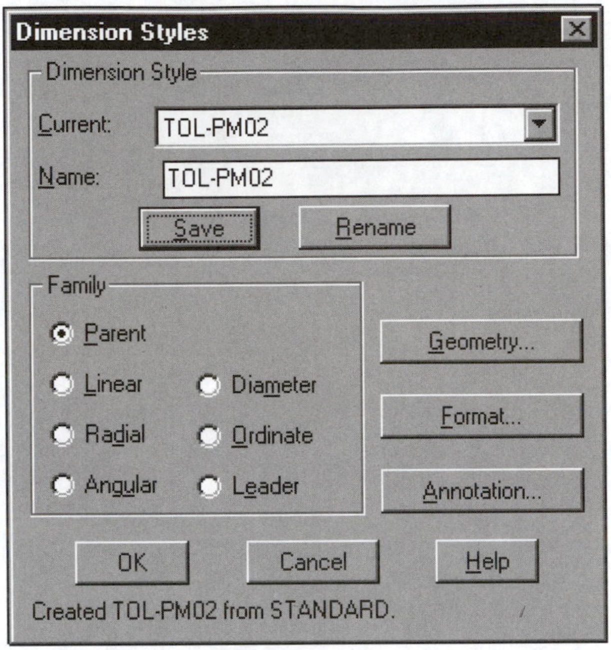

Figure 21-17: Creating a new dimension style

Style Families

Creating a new style based on a previous style lets you design a dimension style and then specify variations on it for the dimension types listed in the Family area of the Dimension Styles dialog box. This option gives you the capability of creating a general setting in the parent style, and variations in the other family styles. The drawing shown in the following figure contains dimensions placed using the same dimension style, yet the number of decimal places for linear is different from radial, and angular. To do this, you set two decimal places for Linear dimensions, one decimal place for Radius dimensions, and zero decimal places for Angular decimal dimensions.

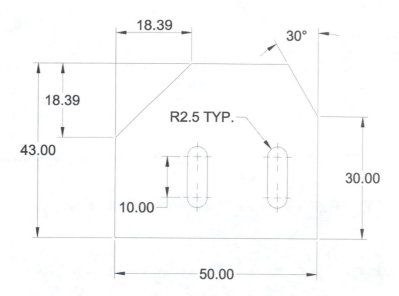

Figure 21-18: Dimensions placed using the same style family

Exercise 21-2: Using Dimension Styles

At the start of the dimensioning process you should consider what dimension styles are required for the drawing. If you are working on a project that already has the styles included in the template file then you simply select the correct style. If that is not the case then you need to create new styles.

In this exercise you will create a new style based on STANDARD. You will also utilize the Family area in the Dimension Styles dialog box to modify the settings for the linear and angular dimensions in STANDARD.

Creating a New Dimension Style

1.   From the File menu, choose Open. Select *dimens3.dwg*, then choose Open. Your drawing should look like the following figure:

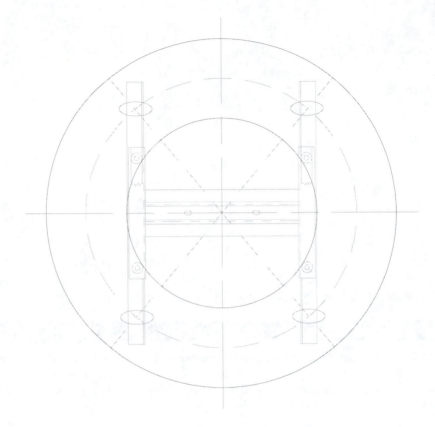

Figure 21-19: The dimens3 drawing

2.   From the Dimension menu, choose Style. Create a new dimension style based on STANDARD by entering **lim-pm02** in the Name area. Then choose Save.

The dialog box should look like the following figure:

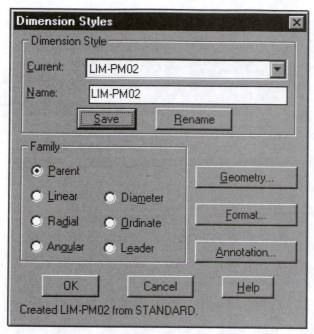

Figure 21-20: The new style name

This style will be used to create tolerance dimensions with a plus and minus value of .02. It is recommended that you create style names that convey a general sense of where the style would be used. In this case the style name indicates a limits dimension (lim), with a plus and minus (pm) value of .02 (02).

3. Select Linear in the Family area. The following settings will apply to linear dimensions, all others will use the settings currently set in Parent.

4. Choose the Geometry button and select the settings as shown in the following figure:

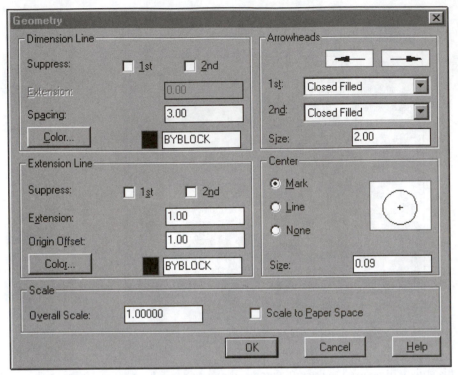

Figure 21-21: The Geometry settings

5. Then choose OK

6. Choose the Format button and select the settings as shown in the following figure:

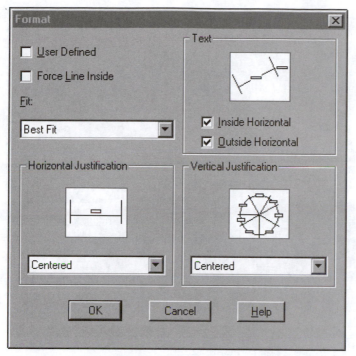

Figure 21-22: The Format settings

7. Then choose OK

8. Choose the Annotation button and select the settings as shown in the following figure:

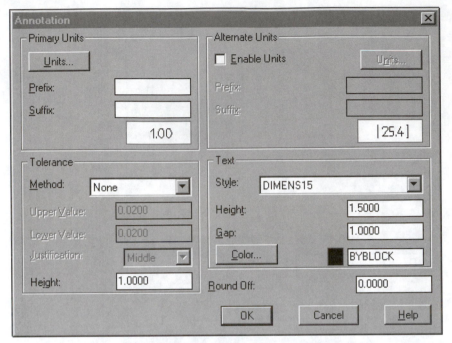

Figure 21-23: The Annotation settings

9. Choose Units and select the settings as shown in the following figure:

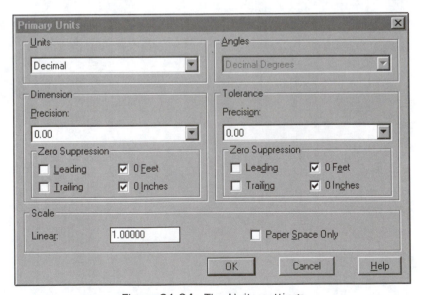

Figure 21-24: The Units settings

10. Then choose OK to return to the Annotation dialog box.

11. Select the Limits from the Method drop-down list, and enter the Upper and Lower Values as shown in the following figure:

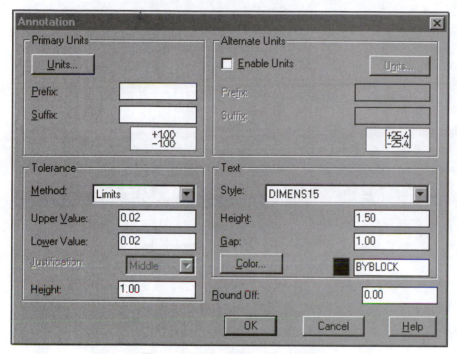

Figure 21-25: Setting the limits

12. Choose OK. Then select Save. The current settings are saved to LIM-PM02.

Creating a Dimension Style for Linear Dimensions

1. With LIM-PM02 as the current style, create and save a new style named **general**. This style will be use for general dimensioning.

2. With Parent selected, set linear to 2 decimal places of precision, and angular to 0 places of precision in the Primary Units section of Annotation. Then choose OK twice.

3. Select Linear from the Family section.

4. Choose the Annotation button and select the settings as shown in the following figure:

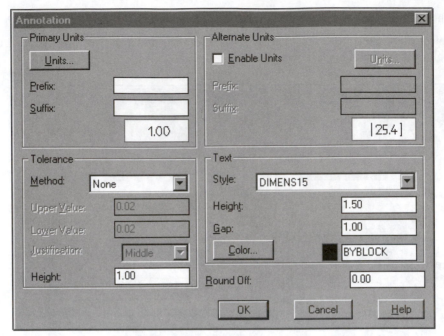

Figure 21-26: The Annotation settings

5. Then choose OK.

6. Select Save. The current settings are saved to GENERAL.

Creating a dimension style for Angular Dimensions

1. Select Angular from the Family section.

2. The settings in Geometry and Format for Angular are the same as the Linear settings created in the previous steps.

3. Choose the Annotation button, then choose Units.

4. Select the settings as shown in the following figure:

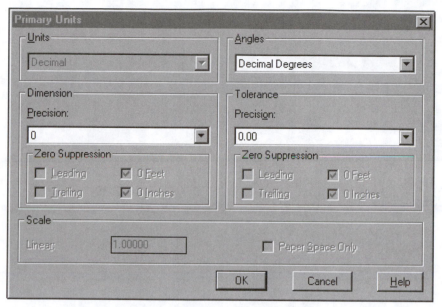

Figure 21-27: The Units settings

5. Choose OK twice to return to the Dimension Styles dialog box. Then choose Save.

6. Choose OK to exit the dialog box.

7. You have created 2 new dimension styles. To use the correct style choose the DDIM command, then select the appropriate style from the Current drop-down list.

8. Complete the drawing, as shown in the following figure, using the dimensioning techniques discussed in this chapter.

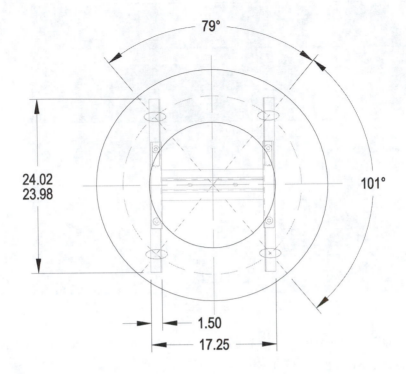

Figure 21-28: The completed drawing

Conclusion

After completing this chapter, you have learned the following:

▶ Dimensions can be edited in a number of ways using
   DDMODIFY, DIMEDIT, and DIMTEDIT.

▶ You can create new styles and use the Family feature in the
   Dimension Style dialog box.

Chapter 22

Printing and Plotting

In this chapter, you learn how to identify the printing and plotting capabilities and features of AutoCAD Release 14. You will improve the appearance and efficiency of your final plotted output by learning how to use the Print/Plot Configuration dialog box.

About This Chapter

In this chapter, you will do the following:

▶ Learn to use the Print/Plot Configuration dialog box.

▶ Select a printer from the Device and Default selection dialog box.

▶ Select a paper size from the Paper Size dialog box.

▶ Set a plot scale and rotation value.

▶ Observe the difference between partial and full plot previews.

▶ Observe the difference between extents and limits with the Full Preview plot feature before plotting.

▶ Create a plot file on the hard disk with the plot to file feature.

Setting Up a Plot

AutoCAD displays drawings in two ways, as *hard copies* or *soft copies*. A *soft copy* is the view of your drawing shown in the drawing window. AutoCAD uses pixels to convert data into the images seen on your monitor. A *hard copy* is the plotted or printed version of your drawing on paper. AutoCAD produces hard copies by converting data into a raster or vector image that is read by a printer or plotter. Hard copies are generally used for finalized drawings, or for drawing reviews. Hard copies can be scaled, unscaled, or realife representations of your work.

When you create a drawing, you should keep the appearance of your finalized drawing in mind. Before your drawing is plotted, you must check all of the plot parameters to make sure your drawing will be plotted correctly. The PLOT command accesses the Print/Plot Configuration dialog box, which is used to set parameters and create plotted drawings. This chapter describes the options associated with the Print/Plot Configuration dialog box.

Methods for opening the Print/Plot Configuration dialog box include:

▶ **Toolbar:** Standard

▶ **Menu:** File > Print

▶ **Command:** PLOT

The Print/Plot Configuration dialog box is shown in the following figure:

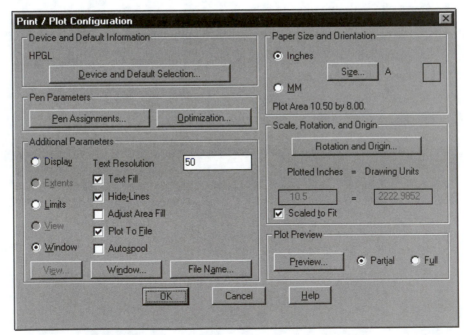

Figure 22-1: Print/Plot Configuration dialog box

Device and Default Selection

The Device and Default Selection option lets you review current information on the configuration of plotters and printers. When you select the Device and Default Selection button, the Device and Default Selection dialog box is displayed. This section describes the uses of the Select a Device, and the Device Specific Configuration areas found in the dialog box. You cannot add a device to the dialog box at this stage. You must reconfigure AutoCAD to recognize another plotter.

The dialog box is shown in the following figure:

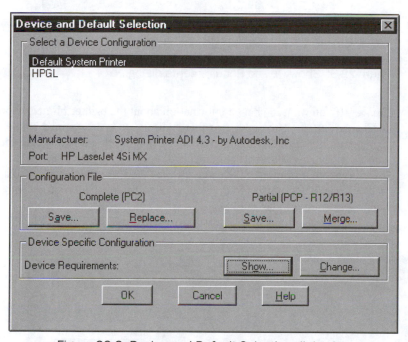

Figure 22-2: Device and Default Selection dialog box

The Select a Device Configuration area displays the names of printing and plotting devices that can be accessed by your computer. The current printing device is always highlighted. To make another device current use the left mouse button to choose the new printer or plotter. When a printing device is made current, information about the manufacturer is displayed at the Manufacturer line. The information displayed at the Port line represents the external path that connects your computer to the printer/plotter.

The Show and Change options in the Device Specific Configuration area let you review or change the printer/plotter settings. A description of these options follows:

> ▶ *Show* - When the Show button is chosen, the Show Device Requirements
> dialog box is displayed with information about the current printing device.
> This dialog box is shown in the following figure:

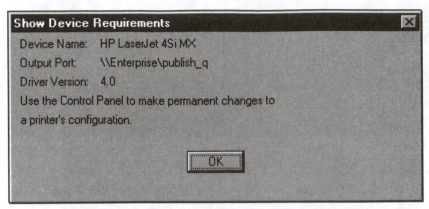

Figure 22-3: Show Device Requirements dialog box

▶ *Change* - If you want to change information about the printer/plotter, choose
   the Change button. The dialog box that is displayed is determined by the type
   of printer/plotter currently being used. For example, if you choose the Change
   button, the Change Device Requirements dialog box may be displayed.
   However, another printing device may display the Print Setup dialog box.
   Each of these dialog boxes require you to enter different information to
   change the printer/plotter settings.

The Change Device Requirements dialog boxes are shown in the following figure:

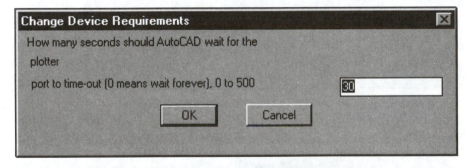

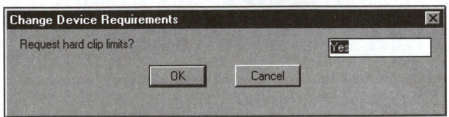

Figure 22-4: Change Device Requirements dialog boxes

The Print Setup dialog box is shown in the following figure:

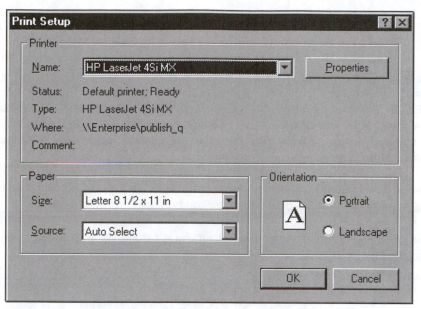

Figure 22-5: Print Setup dialog box

Pen Parameters

When you create a drawing manually, different line widths are used to distinguish items within your drawing and provide contrast. AutoCAD lets you assign a line width to polylines and traces. To assign line widths to different objects, create them on a layer with a specific color and use the Pen assignments options to set a line width for all objects that use that color. AutoCAD distinguishes objects by assigning each object a color. Each color can then be plotted using different pens, linetypes, speeds, and pen widths.

The Pen Parameters area of the Print/Plot Configuration dialog box lets you change pen parameter settings. The Pen Assignments button, and the Optimization button are located in this area. Choosing the Pen Assignments button opens the Pen Assignments dialog box containing the following options:

> ▶ *Color* - Displays the AutoCAD color to which you assign a width, pen, speed, or linetype. If you have a single pen plotter, and have chosen the option to plot different colors with different colored pens, AutoCAD pauses when necessary during the plot and issues a prompt to let you stop and change the pen.

> ▶ *Pen* - Pens are used with pen plotters. This option lets you assign a color to a pen number.

> ▶ *Ltype* - Displays the linetype number assigned to the current color. To see the available plotter linetypes, check the Feature Legend button.

> ▶ *Speed* - This option is used to assign plotting speeds to pens used with pen

plotters. Each plotting speed is assigned a color.

▶ *Width* - The Width option displays the line width assigned to a color. This option generally determines the line width of lines drawn with raster printers.

To set the color, pen number, linetype, speed, or pen width, select the row in the list box that needs modifications. You can change the pen parameters for one row by using the left mouse button to select the row. You can also hold down the SHIFT key and use the left mouse button to select multiple rows. After you choose a row(s), select the specific field from the Modify Values area, then enter your changes. When you change an item in the dialog box it automatically updates the adjacent list box.

Note: If a wide polyline is the same color as other lines, AutoCAD overrides the predetermined setting to properly draw the polyline figure you created.

Be careful when adjusting pen widths. If you get these settings wrong, wide polyline hatched areas, and some linetypes will not print correctly.

The Pen Assignments dialog box is shown in the following figure:

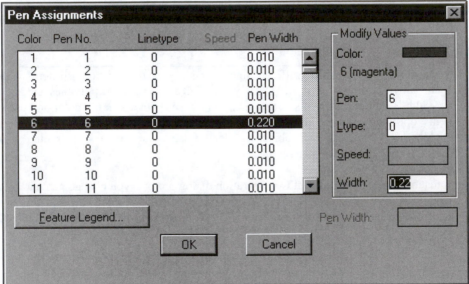

Figure 22-6: Pen Assignments dialog box

The Optimization button located in the Pen Parameters area, displays the Optimizing Pen Motion dialog box. This dialog box has a list of check boxes that increase optimization, by minimizing wasted pen motion and reducing plotting time. By default AutoCAD minimizes pen motion when a drawing is plotted. With the exception of the No Optimization button, the more buttons that are checked the higher the optimization. The printer or plotter type determines if the options in the dialog box are available. The Optimizing Pen Motion dialog box is displayed in the following figure:

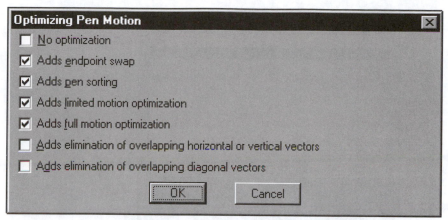

Figure 22-7: Optimizing Pen Motion dialog box

Additional Parameters

The Additional Parameters area of the Print/Plot Configuration dialog box has a list of options used to properly setup your drawing for printing and plotting. The following list describes each of these options:

> ▶ *Display* - This option prints or plots everything shown in the current view of the drawing window.

> ▶ *Extents* - This option plots the area of the drawing that contains objects. Before you print or plot a drawing, use the Zoom to Drawing Extents option to make sure you include all the objects you have created.

> ▶ *Limits* - This option prints or plots everything located inside of the established drawing limits.

> ▶ *View* - This option lets you plot an existing named view. When you select the View button the View Name dialog box is displayed. Select the named view you want to plot, then choose OK. The Print/Plot Configuration dialog box is redisplayed with the view check box highlighted. The View Name dialog box is shown in the following figure:

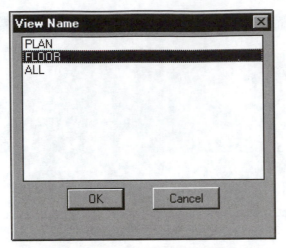

Figure 22-8: View Name dialog box

> ▶ *Window* - This option lets you specify the rectangular area you want to plot, print, or save to plot files. When this button is selected, the Window Selection dialog box is displayed. You can enter coordinates for the First Corner and Other Corner in the dialog box, or use the Pick button to define a window in the drawing. After you specify the plot area choose the OK button. The Window checkbox is now checked. The Window Selection dialog box is shown in the following figure:

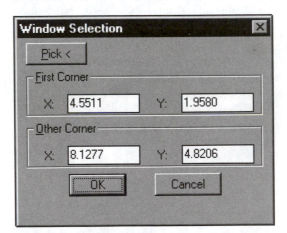

Figure 22-9: Window Selection dialog box

> ▶ *Text Resolution* - This option sets the resolution value for printed text. Lower values increase the plotting speed, but decrease the resolution. Higher values decrease the plotting speed, and increase the resolution. This also sets the resolution, in dots-per-inch, of True Type fonts while plotting. This value is

stored in the TEXTQLTY system variable.

▶ *Text Fill* - If selected, this option displays some text as a solid or filled objects. If the box is unchecked the letters are plotted in an outline form.

▶ *Hide-Lines* - This option if checked, plots model space views with hidden lines removed when a drawing is plotted.

▶ *Adjust Area Fill* - This option lets you compensate for pen-width when plotting wide polylines, solid-filled traces or filled 2D solids. AutoCAD adjusts the boundary of filled areas inward by half a pen width. This can be important for exacting applications like printed circuit board artwork that require greater accuracy.

▶ *Autospool* - This option lets you send a plot file to a printing device while you continue to work. For more information check the Online Help.

▶ *Plot to File* - The Plot to File option lets you create a plot file. Many applications such as word processors can include AutoCAD plot files as illustrations. Instead of printing or plotting your drawing, you have the option of generating files with *.plt* extensions which can be sent to a plotting device, and stored in different directories. The default *.plt* file name is the drawing name. When this button is checked the File Name button is activated.

▶ *File Name* - When the File Name button is selected, the Create Plot File dialog box is displayed. This dialog box lets you name the plot file, then save it to a specified directory. After you assign a name and directory, choose the Save button. The Create Plot File dialog box is displayed in the following figure:

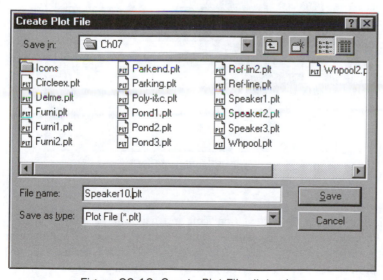

Figure 22-10: Create Plot File dialog box

Paper Size and Orientation

The Paper Size and Orientation area is used to determine the size of the paper that will be used for plotting. You can also select the plot specification units. This is done by choosing Inches or the MM button for millimeters. The Paper Size and Orientation area also have a Size button, Orientation icon, and Plot Area line. These features are discussed in the following section.

Size

The type of printing device that you use determines if the Size button is activated. If the box is not shaded you can select the Size button. The Paper Size dialog box is then displayed. You will find a list of standard paper sizes in a window on the left side of the dialog box. To change the paper size, select one of the rows.

AutoCAD also lets you create your own paper sizes by entering numbers at the User lines. This is done by selecting the Width and Height boxes and entering the desired sizes. To add the new paper sizes to the window, move the cursor inside of the window and press the left mouse button once, or press the ENTER. The new paper size is then displayed. When a paper size is selected the assigned name is displayed next to the Size button in the Paper Size and Orientation area. The Paper Size dialog box is shown in the following figure.

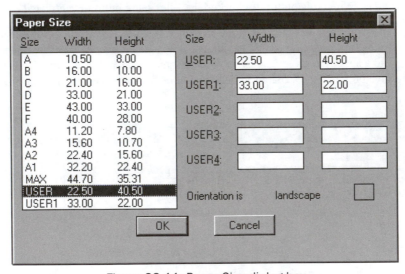

Figure 22-11: Paper Size dialog box

Orientation

When a paper size is selected the Orientation icon and name are displayed in the Paper Size dialog box. The Orientation icon is also shown in the Paper Size and Orientation area. The icon changes depending on the plotting device configuration. The orientation can be Landscape, which means in a horizontal position, or Portrait which is a vertical position.

Plot Area

This line displays the numbers used to create the current paper size. When you are plotting, the paper size (plot area) dimensions used by AutoCAD and the printing device may not be consistent. This can result in plotted drawings that do not fit on the paper. For example HP plotters have wide margins which often cut into the effective plot area.

Scale, Rotation, and Origin

The Scale, Rotation, and Origin area have options that are used to help you plot drawings that meet certain specifications. The plot image is always aligned with the lower left corner of a specified plot area, even when plot is rotated. When a plot is offset, the offset is applied to the entire plot image, rather than the plot origin point. The plot image is offset away from the lower left corner on the page. A description of the options follows.

Rotation

If you select the Rotation and Origin button, the Plot Rotation and Origin dialog box is displayed. The Plot Rotation area of this dialog box lets you select the rotation angle of the plotted drawing. The rotation angle can be set at 0, 90, 180, 270 degrees.

Origin

The Plot Origin area, of the Plot Rotation and Origin dialog box, lets you change the origin of the plot. Generally all drawings are printed from the 0,0 origin located in the lower left corner of the paper. If you want your plot to start in a new location, enter the new coordinate values at the **X** origin and **Y** origin edit boxes. This lets you place multiple plots on the same sheet. The Plot Rotation and Origin dialog box is shown in the following figure:

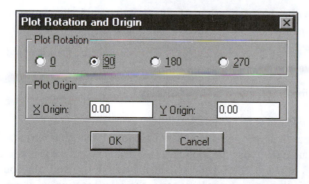

Figure 22-12: The Plot Rotation and Origin dialog box

Scale

Scale factors are used to print or plot text, dimensions, tick marks, arrows, and drawings at the proper size. After you determine your scale factor, AutoCAD lets you enter a scale in the Plotted

Inches = Drawing Units and the Plotted MM = Drawing Units edit boxes. To use this option, make sure the Scale to Fit button is not selected.

Scaled to Fit

If you do not have a scale factor, select the Scale to Fit option. AutoCAD then adjusts the drawing to fit inside of the selected paper size boundaries. This is done by calculating the ratio between the width and height of your specified drawing area and the width and height of the plotted area. The scale factor set by AutoCAD will be displayed in the Plotted Inches = Drawing Units and the Plotted MM = Drawing Units edit boxes.

Previewing a Plot

The Plot Preview area lets you review your drawing before it is printed or plotted. You can select a Partial or Full preview by selecting an option button then choosing Preview. These options are described in the following sections.

> Note You should always run a preview before you plot; this can save paper and time.

Partial

When the Partial button is selected the Preview Effective Plotting Area dialog box is displayed. This dialog box lets you see your drawing in relation to the current paper size. The red outline shown at the top of the dialog box, represents the paper size. The paper dimensions are found below at the Paper Size line. The blue outline represents your drawing. The dimensions of your drawing are located at the Effective area line. If the paper size and effective area are the same dimensions, the outline will have a red and blue dashed line around it.

If there are problems with the orientation of the drawing and the paper size, a warning is displayed at the bottom of the Preview Effective Plotting area dialog box. The most commonly displayed commands are:

- ▶ Effective area too small

- ▶ Origin forced effective area off display

- ▶ Plotting area exceeds paper maximum

If any of these items are displayed, you may need to adjust the plot settings then preview the drawing again. This ensures that your drawing is set up properly before it is printed or plotted.
The dialog box also has a *Rotation icon* located inside of the red and blue outline areas of the dialog box. To determine the rotation angle observe the icon location. Each angle is assigned a location which includes: 0 bottom-left corner, 90 top-left corner, 180 top-right, and 270 bottom-right.

The Preview Effective Plotting Area dialog box is displayed in the following figure:

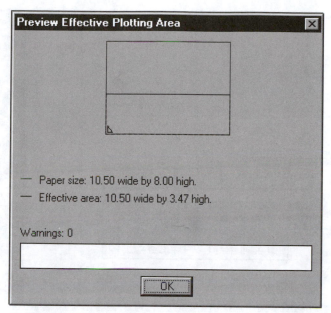

Figure 22-13: Preview Effective Plotting Area dialog box

Full

The Full preview option lets you see the drawing as it would be displayed on paper. When this option is selected a 0-100% meter is displayed at the bottom of the Print/Plot Configuration dialog box. When the regeneration is complete the drawing is displayed, while a regeneration takes place and the PLOT command processes the data. To end the Full preview option, press ESC , ENTER, or the right mouse button to activate the Realtime PAN and ZOOM menu. This menu lets you use the PAN and ZOOM commands to change the view or location of the plotted image on the display screen. The following figure demonstrates how a drawing would be displayed after a Full preview.

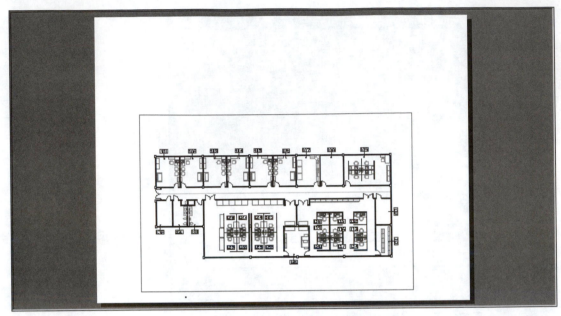

Figure 22-14: Full plot preview

Exercise 22-1: Plotting Drawings

The Print/Plot Configuration dialog box has many options. In this exercise, you use these options to: add a printer driver, choose a printer, set line widths, select a paper size, set the scale, rotation, and origin, define and preview a plot area, and print to a file.

Adding a Print Driver

1.   Open the file *plotme.dwg*. The drawing looks like the following figure:

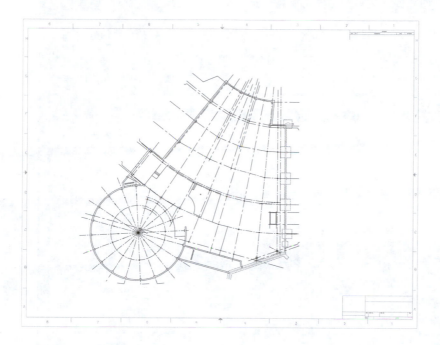

Figure 22-15: Plotme.dwg

2.   In order to plot this drawing, you need to add a new printer driver.

From the Tools menu, choose Preferences. The Preferences dialog box is displayed. Choose the Printer tab, as shown in the following figure:

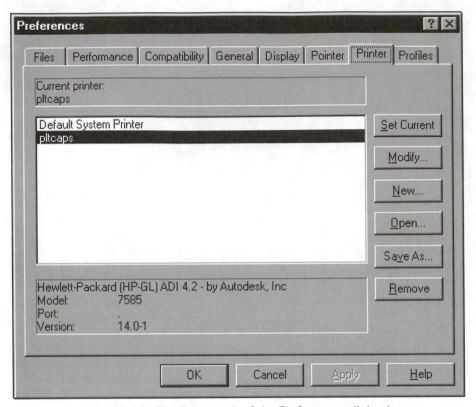

Figure 22-16: The Printer tab of the Preferences dialog box

3. Choose the New button. The Add a Printer dialog box is displayed.

4. Select Hewlett-Packard HPGL/2 devices, ADI 4.3 - for Autodesk by HP.

5. In the Add a description field, enter **HP DesignJet 750C**. Choose OK. The AutoCAD Text Window is displayed.

6. Press ENTER once to continue for supported models. At the `Enter selection:` Command prompt, enter **3** (HP DesignJet 750C).

7. Press ENTER to accept the default to all configuration questions. The Preferences dialog box with the active printer tab is displayed again.

8. In the list of available printers, select HP DesignJet 750C.

9. Choose the Set Current button. The HP DesignJet 750C printer has been added and made to be the current printer. Choose OK.

Choosing a Printer

1. From the File menu choose Print. The Print/Plot Configuration dialog box is displayed, as shown in the following figure:

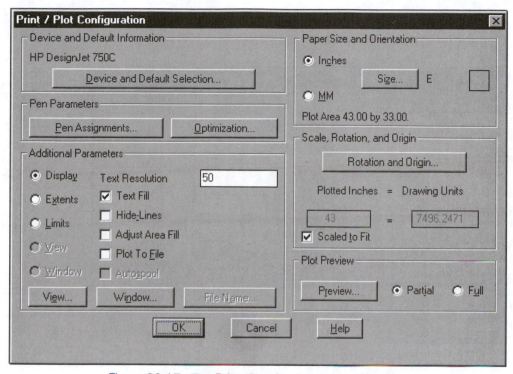

Figure 22-17: The Print/Plot Configuration dialog box

2. In the Device and Default Information area, choose the Device and Default Selection button. The Device and Default Selection dialog box is displayed.

3. In the Select a Device Configuration list box, the device that is highlighted is the current printer. Verify that HP DesignJet 750C is highlighted and choose OK. The Print/Plot Configuration dialog box is displayed again.

Setting Line Widths

1. In the Print/Plot Configuration dialog box, in the Pen Parameters area, choose the Pen Assignments button. The Pen Assignments dialog box is displayed.

2. In the Pen Assignments list, select the Color 1 row.

Notice that in the Modify Values area, pen assignment values for color 1 are now displayed. Notice that the current pen width is .010.

3. You now change the pen width for objects with a color of 1. The color of 1 is red.

In the Modify Values area, in the Width field, enter **.007**. Choose OK.

Selecting a Paper Size

1. In the Print/Plot Configuration dialog box, in the Paper Size and Orientation area, choose the Size button. The Paper Size dialog box is displayed.

2.   In the list of paper sizes, select size E, as shown in the following figure. Choose OK.

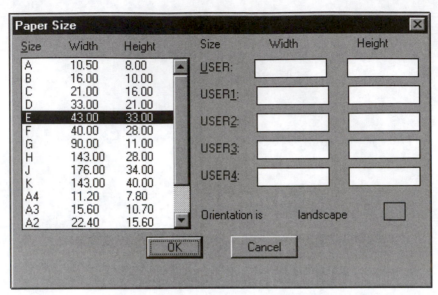

Figure 22-18: The Paper Size dialog box

Setting a Scale, Rotation, and Origin

1.   In the Print/Plot Configuration dialog box, in the Scale, Rotation, and Origin area, choose the Rotation and Origin button. The Plot Rotation and Origin dialog box is displayed.

2.   In the Plot Rotation area, verify that the 0 option is selected.

3.   In the Plot Origin area, verify that the X and Y origin values are set to 0.00. The completed Rotation and Origin dialog box is displayed, as shown in the following figure. Choose OK.

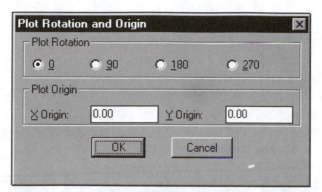

Figure 22-19: The Plot Rotation and Origin dialog box

4.   In the Scale, Rotation, and Origin area, verify that the Scaled to Fit checkbox is checked.

Defining and Previewing the Area to Plot

1.   In the Print/Plot Configuration dialog box, in the Additional Parameters area, select the Extents option.

2.   In the Plot Preview area, select the Full option. Then choose the Preview button. A preview of how the drawing fits on the selected paper is displayed.

3.   After previewing the drawing, press ENTER to continue. The Print/Plot Configuration dialog box is displayed again.

4.   In the Additional Parameters area, select the Limits option.

5.   In the Plot Preview area, verify that the Full option is selected. Then choose the Preview button. A preview of how the drawing fits on the selected paper is displayed.

6.   After previewing the drawing, press ENTER to continue. The Print/Plot Configuration dialog box is displayed again.

Printing to File

1.   Before creating a plot file, you need to change the scale from Scale to Fit to an exact scale. For this plot session, the scale is 1=120 units.

      In the Print/Plot Configuration dialog box, in the Scale, Rotation, and Origin area, clear the Scale to Fit checkbox.

2.   In the Plotted Inches field, enter **1**. In the Drawing Units field, enter **120**.

3.   Before printing the file, you need to verify that the area and scale parameters are set properly by performing a plot preview.

      In the Plot Preview area, verify that the Full option is selected. Then choose the Preview button. When the view is complete, press ENTER.

4.   For this print session you need to create a plot file on disk.

In the Print/Plot Configuration dialog box, in the Additional Parameters area, check the Plot to File checkbox. Then choose the File Name button. The Create Plot File dialog box is displayed.

5. Select the *c:\aotc\level_1* directory. Notice that the plot file name is the same as the drawing name, except that the extension is *.plt*, which is the AutoCAD file extension for plot files.

6. Choose the Save button. This completes this exercise, the completed Print/Plot Configuration dialog box is displayed, as shown in the following figure:

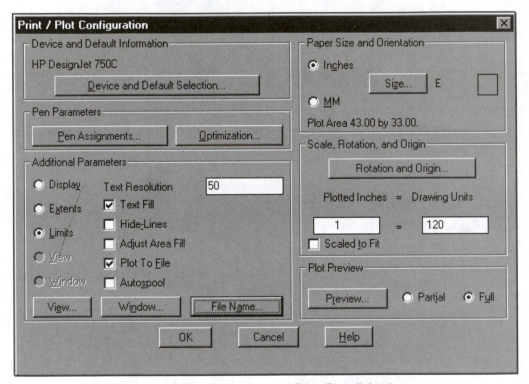

Figure 22-20: The completed Print/Plot dialog box

The exercise is complete, do not save the drawing.

Conclusion

After completing this chapter, you have learned the following:

> ◗ The PLOT command is used to produce hard copies or generate plot files.

> ◗ The Print/Plot Configuration dialog box is used to set plot parameters and created plotted drawings.

> ◗ The Device and Default selection dialog box lets you select a printer, and

review current information on the configuration of plotters and printers.

- The Pen Assignments dialog box is used to change pen parameter settings .

- The Additional Parameters area of the Print/Plot configuration dialog box has a list of options that let you setup your drawing for printing. These include: Display, Extents, Limits, View, Window, Text Resolution, Text Fill, Hide Lines, Adjust Area Fill, Autospool, Plot to File, and File Name.

- The Paper Size dialog box is opened by selecting the Size button. This dialog box is used to select a paper size that will be used for plotting.

- You learned how to use the Scale, Rotation, and Origin area options to set a plot scale and rotation value.

- You learned how to display and the difference between a partial and full preview.

Chapter 23

Inserting Symbols in a Drawing

In this chapter, you learn how blocks let you combine, organize and manipulate several objects as a single object. You also learn how blocks are used to reduce drawing file size.

About This Chapter

In this chapter, you will do the following:

- ▶ Explore the role Blocks play in a drawing.

- ▶ Use Block creation techniques.

- ▶ Insert Blocks in a drawing.

- ▶ Use the Insert dialog box.

- ▶ Scale Blocks.

- ▶ Update Block definitions.

- ▶ Export Block definitions.

- ▶ Explore External References.

Description of Blocks

A block is created from a selection set of one or more objects. You associate the objects together to form a single object, or block definition. Block definitions can be inserted, scaled and rotated within the current drawing.

Block definitions become block references when they are placed within the drawing. Modify commands such as COPY and MOVE treat a block reference as a single object. You can explode a block reference into its individual component objects. These can then be modified as individual objects and redefined back into the block definition.

Blocks reduce drawing file size over using multiple copies of individual objects. With multiple references of a block, AutoCAD stores just enough data to point to the original block definition in the drawing database.

Creating Block Definitions

The BLOCK and BMAKE commands are used to make block definitions. You must perform the following steps before you can use these commands:

▶ Create the objects that are intended to be part of a block definition and put them in place.

▶ Determine a name for the block.

▶ Determine an insertion point that will be referenced during block insertion.

Using the BLOCK Command

After completing the preliminary steps, select one of the following methods to invoke the BLOCK or BMAKE commands:

▶ **Toolbar:** Draw

▶ **Menu:** Draw > Block > Make

▶ **Commands:** BLOCK, BMAKE

All Block command selections except BLOCK, display the Block Definition dialog box, as shown in the following figure:

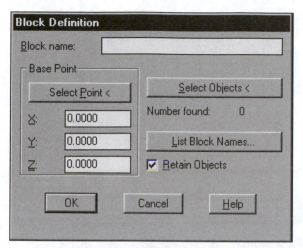

Figure 23- 1: Block Definition dialog box

The Block Definition Dialog Box

The following list describes the items in the Block Definition dialog box:

- *Block name* - Enter the name of a block in this field. Block names can be up to 31 characters in length, and can contain letters, digits, special characters, the dollar sign ($), the hyphen (-), and the underscore (_). AutoCAD converts all letters to uppercase.

- *Base Point* - The base point specified is used as the reference base point for insertions of the block. The base point is also the point during insertion, about which the block can be rotated, and from which the block can be scaled.

- Select Objects - Specifies the objects to include in the new block. Use any object selection method. To complete the object selection process, press ENTER, and AutoCAD redisplays the dialog box. The total number of objects selected is displayed.

- *List Block Names* - Displays the Block Names in This Drawing dialog box. You can enter wild-card characters to query the list in the Pattern field, as shown in the following figure:.

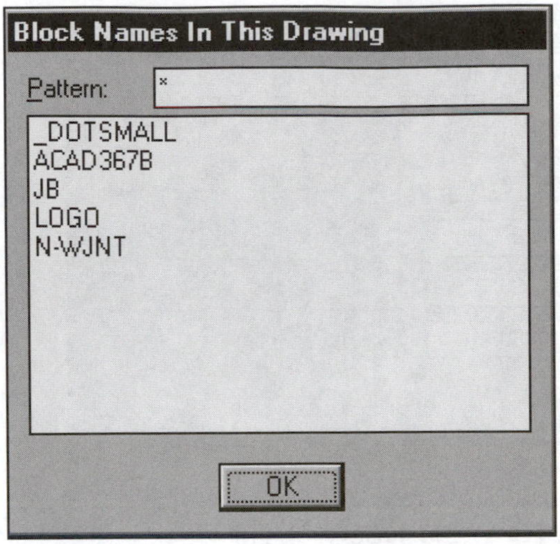

Block Names In This Drawing

Pattern: [*]

```
_DOTSMALL
ACAD367B
JB
LOGO
N-WJNT
```

[OK]

Figure 23-2: Listing of block names In this drawing

▶ *Retain Objects* - Retains objects in your drawing that you have selected to be in the block.

Note: The objects you select are deleted from the drawing unless you selected Retain Objects or enter *oops* at the Command prompt after completing the BLOCK command.

After the required information is defined in the dialog box and the objects have been selected,choose OK or press ENTER to complete the block definition creation.

Note: If you enter the name of an existing block, AutoCAD displays the following Warning. By redefining a block, you automatically update all references to that block. The Warning is shown in the following figure:

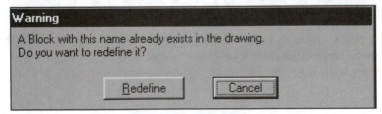

Warning

A Block with this name already exists in the drawing.
Do you want to redefine it?

[Redefine]   [Cancel]

Figure 23- 3: Warning for redefining an existing block definition

You can also create block definitions at the Command prompt. Command prompt input for the BLOCK command looks like the following:

```
Command: block
Block name (or ?): new_block_name
```

Insertion base point: **0,0**

Select objects: Other corner: 6 found

Select objects: ENTER

> Note: If you enter the name of an existing block, AutoCAD prompts as follows:
>
> *Block "name" already exists.*
>
> *Redefine it? <N>* Enter y or n, or press ENTER
>
> By redefining a block, you automatically update all references to that block.

By entering a **?** at the Block name prompt, you can display a list of all previously-defined blocks in the AutoCAD Text Window, as shown in the following example:

Command: **block**

Block name (or ?): **?** or List previously defined blocks

Block(s) to list <*>: ENTER

Defined blocks.

  _DOTSMALL

  ACAD367B

  JB

  LOGO

  N-WJNT

User	External	Dependent	Unnamed
Blocks	References	Blocks	Blocks
5	0	0	4

The following information is included in the list:

> ▶ *Defined blocks* - AutoCAD lists the block names.
>
> ▶ *User Blocks* - Number of user blocks in the current drawing
>
> ▶ *External References* - Number of external references in the current drawing.
>
> ▶ *Dependent Blocks* - Number of dependent blocks in the current drawing. You cannot insert or select dependent blocks, but you can control the visibility, color, and linetype.
>
> ▶ *Unnamed Blocks* - AutoCAD creates unnamed (also called anonymous) blocks to support hatch patterns, associative dimensioning, and PostScript images imported with PSIN. AutoCAD also creates unnamed blocks for objects that you cannot gain access

Nested Blocks

A block definition can contain other blocks as objects in the selection set. Blocks contained within a block reference are called *nested* blocks. Nested blocks can not be inserted into or used to create blocks that reference themselves. Although block nesting can be useful, floating layers, colors, and linetypes can make nesting complicated if they are not used correctly.

Features of Layer 0, BYLAYER, and BYBLOCK

When you insert a block definition that consists of objects drawn on layer 0 and assigned the color and linetype BYLAYER, it is placed on the current layer. These block references assume the color and linetype properties of the current layer.

You can create blocks definitions from objects that are drawn on different layers with different colors and linetypes. This method is used to preserve the layer, color, and linetype information of objects in the block. Each block reference within the drawing will have its objects on their original layer with their original color and linetype.

A block consisting of objects that have color or linetype specified with BYBLOCK is drawn with the color and linetype that are current when the block is inserted. If the color and linetype are not explicitly assigned, the block assumes the color and linetype of the layer. The following rules apply to setting up layers, colors and linetypes for block definitions:

> ▶ If all references of a particular block need the same layer, color, and linetype properties, assign properties explicitly to all objects in the block (including any nested blocks).

> ▶ If you want to control the color and linetype of each reference of a particular block by using the color and linetype of the layer on which you insert it, draw each of the block's objects (including any nested blocks) on layer 0 with color and linetype set to BYLAYER.

If you want to control the color and linetype of each reference of a particular block using the current explicit color and linetype, draw each of its objects (including any nested blocks) with color and linetype set to BYBLOCK.

Exercise 23-1: Creating Blocks

In order to use blocks in a drawing you must first create them. In this exercise, you define block definitions for two architectural symbols that are inserted into the drawing in Exercise 23-2.

Creating Block Definitions

1. Open the file *block_01.dwg*. If you do not see the view in the following figure, restore the LAV view using the Named Views button on the Standard toolbar. You are going to create a block definition for a bath lavatory sink.

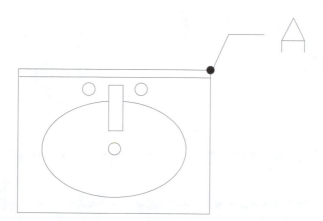

Figure 23-4: Objects used to define a lavatory block symbol

2.   From the Draw toolbar, choose the Make Block button. The Block Definition dialog box is displayed.

3.   In the Block Name field, enter **lav_24x18**. Choose the Select Point button and select a base point at the intersection of A in the previous figure. The Block Definition dialog box is displayed again.

4.   Choose the Select Objects button and use window selection to select the objects in the drawing window that define the lavatory. Then press ENTER to return to the Block Definition dialog box.

5.   Clear the Retain Objects checkbox and choose OK. The objects disappear from the drawing window. The lav_24x18 block has been created.

6.   Restore the Tub view, shown in the following figure. You are now going to create a block definition for a bath tub symbol.

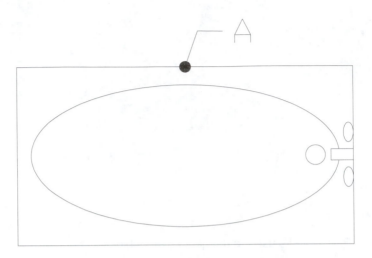

Figure 23-5: Objects for Tub block definition

7. From the Draw toolbar, choose the Make Block button. Enter **tub** in the Block Name field, and select a base point at the midpoint of the line A, as shown in the previous figure. The Block Definition dialog box is redisplayed.

8. Choose the Select Objects button and window the objects in the drawing window that define the bath tub. Then press ENTER to return to the Block Definition dialog box.

9. Clear the Retain Objects checkbox and choose OK. The following Warning dialog box is displayed, as shown in the following figure, informing you that a block by the name of Tub already exists. Choose the Redefine button to update the Tub block.

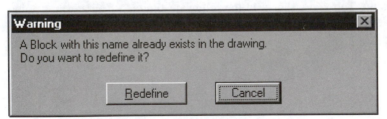

Figure 23-6: Warning for redefining an existing block definition

The objects disappear from the drawing window. The Tub block has been redefined.

10. Restore the Bath_1 view.

11. Save the drawing as *block_02.dwg* so that you can use it for the next exercise.

Inserting Blocks into a Drawing

Once you have created your block definitions, they have to be inserted into the drawing in order to be displayed. The INSERT command is used to place a block definition or an existing drawing into your current drawing.

Using the Insert Dialog Box

Methods for invoking the INSERT command include:

▶ **Toolbar:** Draw

▶ **Menu:** Insert > Block

▶ **Commands:** DDINSERT, INSERT

To open the Insert dialog box, use the toolbar and menu methods. The Insert dialog box is shown in the following figure:

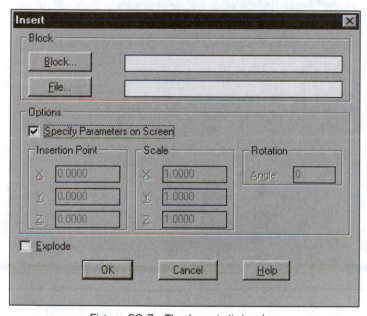

Figure 23-7: The Insert dialog box

The following list describes the options in the Insert dialog box:

▶ *Block* - Specifies the name of a block to be inserted or the name of a file to be inserted as a block. Choosing the Block button displays the Defined Blocks dialog box in figure 14-8.

▶ *File* - Choosing the File button displays the Select Drawing File dialog box.

▶ *Options* - Specifies insertion point, scale, and rotation angle of the block definition being inserted. These values reference the current UCS. By checking the Specify Parameters on Screen checkbox, you instruct AutoCAD to prompt you for the insertion point, scale, and rotation angle at the Command prompt.

▶ *Explode* - Inserts the individual parts of the block. When you check the Explode checkbox, you specify only an X scale factor. Component objects of a block drawn on layer 0 remain on that layer. Objects having color BYBLOCK are white. Objects with linetype BYBLOCK have continuous lines.

The Defined Blocks Dialog Box

Using the Defined Blocks dialog box, you can select an existing block definition by selecting the block name. In the Pattern field, you can enter wild-card characters to query the list. The Defined Blocks dialog box is shown in the following figure:

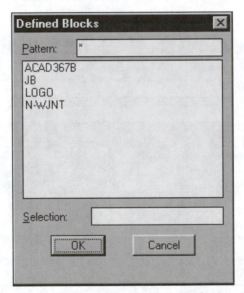

Figure 23-8: Defined Blocks available for insertion

You can also insert block definitions using the Command prompt. Command prompt input for the INSERT command looks like the following:

```
Command: insert
Block name (or ?)<current>: jb
Insertion point: 0,0
X scale factor <1> / Corner / XYZ:   ENTER
Y scale factor (default=X):   ENTER
Rotation angle <0>:   ENTER
```

The following information defines the command input required for the INSERT command:

▶ *Block name (or ?) <current>* - Specify the name of a block to be inserted or the name of a file to be inserted as a block. Entering **?** lists the current block definitions in the drawing. Entering a tilde (~) displays the Select Drawing File dialog box. Preceding the name of the block with an asterisk (*) separates the block's objects during insertion. If you enter a block name without a path name, AutoCAD searches for a currently defined block definition by that name. If no such block definition exists in the current drawing, AutoCAD searches the library path for a file of the same name. If AutoCAD finds such a file, the file name is used for the block name when AutoCAD inserts the block definition.

▶ *Insertion point* - Specify a point or enter an option

▶ *X scale factor <1> / Corner / XYZ* - Enter a value or option, or press ENTER to confirm an X scale factor of 1.

▶ *Y scale factor (default=X)* - Enter a value if different from the X scale value, or press ENTER to default to the value given for X.

▶ *Rotation angle <0>* - Enter a value, or press ENTER to accept the default.

Note: A negative scale factor can be used to mirror the block's insertion, turn the block reference upside down, or both.

Updating Block Definitions

You can use the BLOCK and BMAKE commands to redefine blocks. Redefinition affects references of the block as well as future insertions of a block reference. The process requires the following steps:

▶ Insert the desired block at a location where it can be edited, or use an exiting reference of the block in the drawing.

▶ Explode the block reference to be updated.

▶ Perform any editing tasks that are required for the update.

▶ Select the BLOCK command, and give the same name as the exploded block for the updated block definition.

▶ Complete the BLOCK command as previously described in this chapter. You receive the following message if you entered **block** at the Command prompt:

```
Block "name" already exists.
Redefine it? <N> Enter y or n, or press ENTER
```

▶ The Warning dialog box in the following figure is displayed if you used the BMAKE command. Either way, you must redefine the block for the update to take affect.

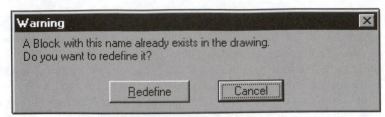

Figure 23-9: Warning for redefining an existing block definition

Exporting Block Definitions

You use the WBLOCK command to create a separate drawing file from a block definition. The drawing file can then be used as a block definition for other drawings. AutoCAD considers any drawing you insert into another drawing to be a block definition.

Using the WBLOCK Command

WBLOCK is accessed at the Command prompt. After you enter the command, the Create Drawing File dialog box is displayed, as shown in the following figure:

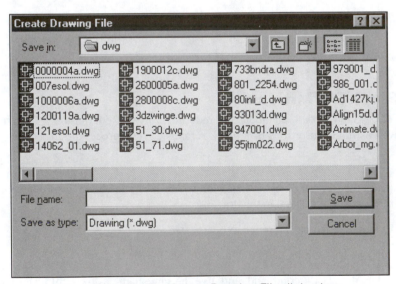

Figure 23-10: The Create Drawing File dialog box

It is prompting you for a drawing file name. If you enter a file name that already exists, then the alert box is displayed, as shown in the following figure. You have to decide if you want to replace an existing drawing file with the block definition.

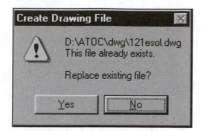

Figure 23-11: Warning to replace exiting file

When you have completed the file naming process, you are returned to the prompt:

Block name:

Enter one of the following options at the Command prompt:

▶ The name of an existing block, which writes that block to a file

▶ An equal sign (=), which specifies that the existing block and the output file have the same name

▶ An asterisk (*), which writes the entire drawing to the new output file, except for unreferenced symbols

▶ Press ENTER. If you press ENTER at the Block name prompt, AutoCAD prompts you for a base point at which to insert the block. Then AutoCAD prompts you to select the objects to write to a file.

Note: In the new drawing, the World Coordinate System (WCS) is set parallel to the UCS in effect at that time.

Exercise 23-2: Using and Updating Blocks

In exercise 1 you created two block definitions. In this exercise, you insert those block definitions, create a new block definition of a bathroom, and insert the bathroom block into a one bedroom apartment floor plan. You also update the bathroom block definition to reflect a design change.

Inserting Blocks

1. Open the file *block_02.dwg* created in exercise 23-1. Restore the Bath_1 view using the Named Views button on the Standard toolbar. The following figure displays this view:

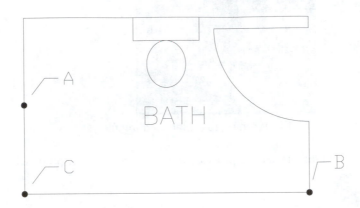

Figure 23-12: Bathroom layout with Bath_1 view restored

2. From the Draw toolbar, select the Insert Block button. The Insert dialog box is displayed.

3. Choose the Block button and choose the Tub block from the list of defined blocks. Choose OK.

4. Check the Specify Parameters on Screen checkbox, and choose OK. You are returned to the Command prompt and the drawing editor.

5. For the Insertion point use the midpoint of line A, as shown in the previous figure.

6. Press ENTER twice through the default X and Y scales, and enter a Rotation angle of **90** to complete the command.

7. From the Draw toolbar, select the Insert Block button. The Insert dialog box is displayed.

8. Choose the Block button and select the Lav_24x18 block from the list of defined blocks. Choose OK.

9. Make sure that the checkbox for Specify Parameters on Screen is checked and choose OK. You are returned to the Command prompt and the drawing editor.

10. For the Insertion point, use the intersection at B, as shown in the previous figure.

11. ENTER twice through the default XY scale, and enter a Rotation angle of **-90**.

   You should have a bathroom layout that looks like the following figure:

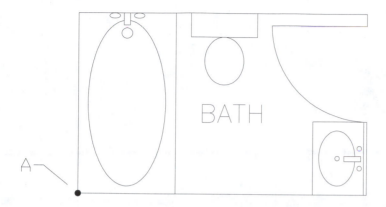

Figure 23-13: Finished bathroom layout

Creating and Inserting a New Block Definition

1.   From the Draw toolbar, choose the Make Block button.

2.   Enter **bath** in the Block Name field, and choose the Select Point button. Define a base point at the intersection of A in the previous figure. The Block Definition dialog box is redisplayed.

3.   Choose the Select Objects button and window the objects in the drawing window that define the bath, including the tub and lav_24x18 block references. Then press ENTER to return to the Block Definition dialog box.

4.   Clear the Retain Objects checkbox, and choose OK.

5.   Restore the Unit_plan view to see the floor plan in the following figure:

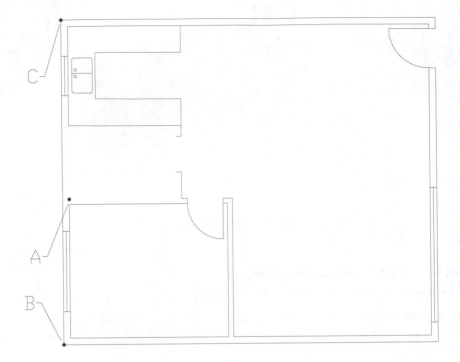

Figure 23-14: Unit floor plan ready for bathroom layout

6. From the Draw toolbar, choose the Insert Block button. The Insert dialog box is displayed.

7. Choose the Block button and select the Bath block from the list of defined blocks. Then choose OK. Make sure that the checkbox for Specify Parameters is checked and choose OK. You are returned to the Command prompt and the drawing editor.

8. For the Insertion point use the node at A, as shown in the previous figure.

9. ENTER three times through the default XY scale and the Rotation angle.

You have placed the bathroom layout into the unit floor plan, as shown in the following figure:

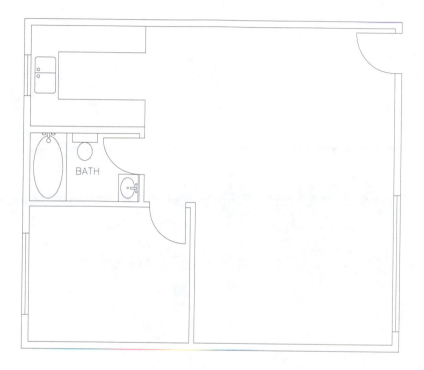

Figure 23-15: Floor plan with bathroom block reference

10. You can continue the floor plan layout process by creating a block definition for the Unit_floor_plan in the previous figure. Use the intersection at B in Figure 14-14 as the insertion base point. Insert the first Unit_floor_plan block definition with an insertion point of **0,0**. Use the intesection at C in Figure 14-14 as the insertion point for additional block definitions, as shown in the following figure:

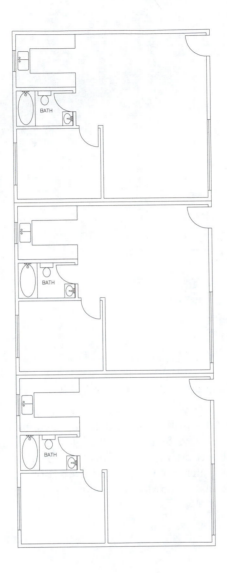

Figure 23-16: Multiple references of the Unit_floor_plan block

Update a Block

1.   Restore the Bath_2 view, as shown in the following figure, using the Named Views button on the Standard toolbar.

Changes have been ordered in the bathroom layout for the apartment units. The door location and swing direction are to be changed. A larger lavatory is to be placed next to the watercloset.

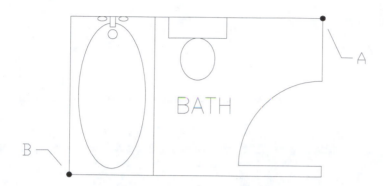

Figure 23-17: Bath_2 view restored

2.   From the Draw toolbar, choose the Insert Block button. The Insert dialog box is displayed.

3.   Choose the Block button and select the Lav_32x18 block from the list of defined blocks. Make sure that the checkbox for Specify Parameters on Screen is checked and choose OK. You are returned to the Command prompt and the drawing editor.

4.   For the Insertion point use the intersection at A in the previous figure.

5.   Press ENTER three times through the default XY scale and Rotation angle.

6.   Enter **block** at the Command prompt. Enter **bath** in the Block Name field and press ENTER. The following message is displayed:

Block "name" already exists.

Redefine it? <N>

7.   Enter **y** and then press ENTER.

8.   For the Insertion base point, select a base point at the intersection at B in the previous figure.

9.   Window to select all of the objects in the Bath_2 view.

10.   You can now restore the Unit_plan view or do a ZOOM Extents to verify that all references of the bath block have been updated to reflect the design changes.

If you created the Unit_floor_plan block reference, you have to explode that block in order to clean up the door opening for the bathroom as seen in the following figure. You can then redefine the Unit_floor_plan block to reflect the change in all of its references.

The following figure shows the completed drawing:

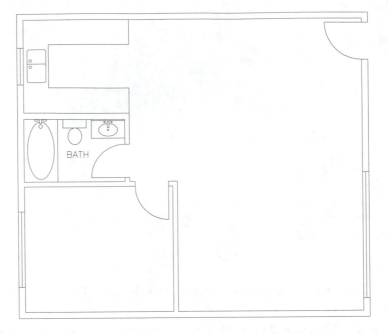

Figure 23-18: Complete single unit floor plan

Overview of External References

External references (xrefs) are a way of combining drawing files. When you externally reference a drawing, you can view and snap to the referenced drawing from your current drawing. However, each drawing's data is still stored and maintained in separate drawing files. A link or association is established between the drawing files.

Conclusion

After completing this chapter, you have learned the following:

- ◗ Using blocks, you can combine, organize, and manipulate several objects as a single object.

- ◗ You use the BLOCK command and the Block Definition dialog box to create block definitions. In order to create block definitions, the objects must first be created.

- ◗ You use the INSERT command and the Insert dialog box to insert a block into your drawing.

- ◗ You can create a new block definition, such as the bathroom, using existing objects, including block references.

▶ You use the BLOCK and BMAKE commands to redefine blocks. Redefinition affects references of the block as well as future insertions of a block reference.

▶ You can use the WBLOCK command to create a separate drawing file from a block definition. The drawing file can then be used as a block definition for insertion into other drawings.

Chapter 24

Using Utility Commands

In this chapter, you learn how to use the PURGE and RENAME commands.

About This Chapter

In this chapter, you will do the following:

- ▶ Reduce file sizes by using the PURGE command.
- ▶ Use the DDRENAME command to list the named objects in a drawing file.
- ▶ Rename named objects with the DDRENAME command.

The PURGE Command

The PURGE command compacts the drawing by eliminating all unused named objects. Named objects include: blocks, dimension styles, layers, line types, shapes, styles, and mlinestyles.

Methods for invoking the PURGE command include:

▶ **Menu:** File > Drawing Utilities

▶ **Command:** PURGE

When the PURGE command is invoked you will have the option of purging ALL, which means you can eliminate every unused named object in your drawing. You also have the option to restrict purging to objects from an individual category.

It is important to keep drawings as small as possible. From an efficiency standpoint, it makes sense that a smaller drawing uses less disk space and less system resources. After determining that items are no longer needed, you can remove them from the drawing database by using the PURGE command. Only unreferenced items can be purged. One example is a block containing objects on multiple layers. You cannot purge the block layer names until after you purge the block. Therefore, when dealing with nested referenced items it is necessary to use the PURGE command multiple times. Performing multiple purges decreases your drawing file size by 10 - 20%. A purge can be performed at anytime during a drawing session.

Renaming Objects

AutoCAD lets you rename objects that are within, or associated with a drawing. When the RENAME command is entered, a list of categories that generally have named objects associated with them is displayed. You have the option of selecting Block; Dimstyle; Layer; LType; Style; UCS; View, or Viewport. After you choose one of these options, you are prompted to enter the name of the old object, then enter the name for the new object.

Methods for invoking the RENAME command include:

▶ **Menu:** Format > Rename

▶ **Command:** RENAME

Instead of entering information at the Command prompt, AutoCAD lets you use the Rename dialog box to achieve the same objective. To access the dialog box enter DDRENAME at the Command prompt, or choose Rename from the Format menu. To rename an object using the Rename dialog box choose a category from the Named Objects window. If the category has named objects in it, they will be displayed in the Items window. To specify the object to be renamed, select the object from the Items window, or enter a name in the Old Name box. Enter the new name in the Rename To box, then select the Rename To button to make the change.
The Rename dialog box is shown in the following figure:

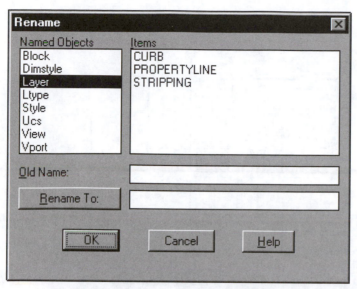

Figure 24-1: Rename dialog box.

Exercise 24-1: Cleaning Up Drawings

In this exercise, you will perform a primary and secondary purge to reduce the drawing file size, view the results of multiple purges, and rename objects.

Primary Purge

1.   Open the file, *cleanup.dwg*. The drawing looks like the following figure:

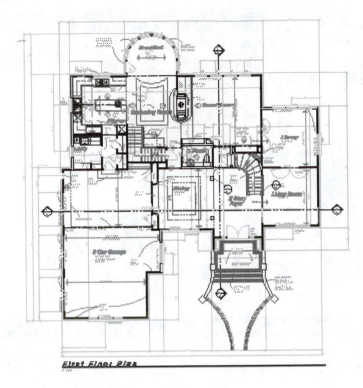

Figure 24-2: Cleanup.dwg

2. You will now save the file with a new name before cleaning it up.

    From the File menu choose Save As. The Save As dialog box is displayed.

3. Enter **cleanup1** in the File Name field. Then choose the Save button.

4. You will now use the PURGE command to reduce the drawing file size by removing unneeded objects from the drawing.

    From the File menu, choose Drawing Utilities. Then choose Purge.

5. In the Purge submenu, you have the option of purging specific groups or all unreferenced objects. Choose the All option.

6. In response to the a Names to purge <*> Command prompt, press ENTER to accept the default command option.

7. In response to the Verify each name to be purged ? <Y> Command prompt, enter **n** and press ENTER.

8. From the File menu, choose Save. A copy of the file has now been saved with the name, *cleanup1.dwg*.

    This file represents the first stage of using the PURGE command. In some drawings not all objects can be purged on the first try and it is sometimes necessary to repeat the PURGE command.

Secondary Purge

1. You will now use the PURGE command again to remove unneeded objects that the first purge could not remove from the drawing.

    At the Command prompt, enter **purge** or its keyboard alias **pu**.

2. In response to the following purge command option prompt, enter **a** for the All option and press ENTER.

    Purge unused Blocks/Dimstyles/LAyers/LTypes/SHapes/STyles/ Mlinestyles/All:

3. In response to the `Names to purge <*>` Command prompt, press ENTER to accept the default command option.

4. Since you want to purge all unreferenced objects and do not want to be prompted for every name, in response to the `Verify each name to be purged? <Y>` Command prompt, enter **n** and press ENTER.

5. You will now save the file with a new name after cleaning it up.

    From the File menu, choose Save As. The Save As dialog box is displayed.

6. Enter **cleanup2** in the File Name field. Then choose the Save button.

    This file represents the secondary stage of using the PURGE command.

7. Use the PURGE command again, following either of the previous purge procedures.

8. After completing the PURGE command, save the file with the name **cleanup3**.

    This file represents the last stage of using the PURGE command.

Viewing the Results of Multiple Purges

1. From the File menu, choose Open. The Select File dialog box is displayed.

2. In the Select File dialog box, List view is the default. To display file information, choose the Details button. The dialog box now displays the file sizes of the cleanup drawings after using multiple purges. Notice that after using multiple purges, the file size has dropped to one-half the original size.

3. Choose Cancel to close the dialog box.

Renaming Objects

1. The primary purpose of the Rename dialog box is for changing object names in drawings. However, it also a very efficient method of identifying object names that reside within specific object item groups for the purpose of determining the structure of an unfamiliar drawing.

From the Format menu, choose Rename. The Rename dialog box is displayed, as shown in the following figure:

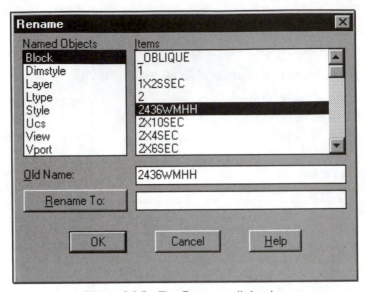

Figure 24-3: The Rename dialog box

2.   In the Rename dialog box, under Named Objects, select Block. All blocks in the drawing are now displayed in the Items list box. Scroll down the list and observe all the blocks in the drawing.

3.   In the Rename dialog box, in the Named Objects list box, select Style. All font names in the drawing are now displayed.

4.   Select Dimtxt. Dimtxt is now displayed in the Old Name box.

5.   In the Rename field, enter **olddimtxt**, then choose the Rename To button. Notice the original Dimtxt font style name is now called Olddimtxt.

6.   Select the Close button.

7.   You have now experienced how multiple purges affect the file size, and how to use the Rename dialog box to change the name of items or identify named objects. You have now finished this exercise. Do not save the drawing.

Conclusion

After completing this chapter, you have learned the following:

> ◗   To invoke the PURGE command, you enter **purge** at the Command prompt, or choose File menu, Drawing Utilities, then Purge.

> ◗   Use the PURGE command to perform multiple purges that reduce file sizes.

◗   The DDRENAME command is used to review drawing object content.

◗   You can change the name of blocks by using the RENAME command to access
    the Rename dialog box. The dialog box lets you choose an item to be
    renamed, then enter a new name.

Index